Genetically engineered proteins represent a completely new generation of drugs which permit the replacement or reinforcement of vital components which may be defective or deficient. However, such drugs must be precise copies of the proteins in the human body that they are intended to replace. Because of their complexity, these molecules can only be made by biotechnological methods, and the only systems able to produce completely accurate copies of the most complex proteins are cultures of animal cells.

This book covers all aspects of the new technologies needed to turn animal cells into an acceptable and cost-effective tool for drug production. These include modifying cells genetically so that they produce the right product in high yield, getting them to grow reproducibly on an industrial scale, and extracting the required product from them. It also covers biological safety issues raised by production in animal cells and the verification of the chemical and biological nature of the protein drug produced.

The work covers the latest developments in all of these areas and how they all must be integrated for the design of an effective biotechnological production process. It therefore provides a comprehensive guide to this area of biotechnology and will be valuable to students, scientists and technicians who are active in biotechnology research, development and industrialization.

Cambridge Studies in Biotechnology

Editors: Sir James Baddiley, N.H. Carey, I.J. Higgins, W.G. Potter

11 Animal cells as bioreactors

11 Animal cells as bioreactors

Other titles in this series

1 The biotechnology of malting and brewing
J.S. Hough
2 Biotechnology and wastewater treatment
C.F. Forster
3 Economic aspects of biotechnology
A.J. Hacking
4 Process development in antibiotic fermentations
C.T. Calam
5 Immobilized cells: principles and applications
J. Tampion and M.D. Tampion
6 Patents: a basic guide to patenting biotechnology
R.S. Crespi
7 Food biotechnology
R. Angold, G. Beech and J. Taggart
8 Monoclonal antibodies in biology and biotechnology
K.C. McCullough and R.E. Spier
9 Biotechnology of microbial exopolysaccarides
I.W. Sutherland
10 Microalgae: biotechnology and microbiology
E.W. Becker

Animal cells as bioreactors

TERENCE CARTWRIGHT
TCS Biologicals, Ltd., U.K.

CAMBRIDGE UNIVERSITY PRESS
Cambridge, New York, Melbourne, Madrid, Cape Town, Singapore, São Paulo, Delhi

Cambridge University Press
The Edinburgh Building, Cambridge CB2 8RU, UK

Published in the United States of America by Cambridge University Press, New York

www.cambridge.org
Information on this title: www.cambridge.org/9780521103107

First published 1994
This digitally printed version 2009

A catalogue record for this publication is available from the British Library

Library of Congress Cataloguing in Publication data

Cartwright, Terence.
Animal cells as bioreactors / Terence Cartwright.
p. cm. – (Cambridge studies in biotechnology : 11)
Includes bibliographical references.
ISBN 0–521–41258–7
1. Animal cell biotechnology. 2. Pharmaceutical biotechnology.
I. Title. II. Series.
615′.19 – dc20 93–33980

ISBN 978-0-521-41258-2 hardback
ISBN 978-0-521-10310-7 paperback

Contents

1 Introducing the Animal Cell as a Bioreactor

Generalities

A bioreactor is essentially a tool or device for generating product using the synthetic or chemical conversion capacity of a biological system. From the 1970s onwards there has been steady increase of attempts to harness biological systems to perform specific and difficult chemical tasks as demands for lower energy consumption and the requirement for ever more intricate syntheses have increased. Bioreactors have ranged from immobilized and engineered enzymes, through an ever increasing number of natural and recombinant microbial systems, to cells derived from animal tissues.

When the aim is to reproduce the complex molecules found in animals, cultured animal cells offer unique qualities as bioreactors because they alone are capable of accurately reproducing effectively the whole of the biological chemistry which operates in our bodies. Thus, they can, in principle, reproduce any molecule occurring there which may find application in medicine. In practice, it is in the production of proteins and their derivatives for diagnostic, prophylactic and therapeutic use that animal cells are currently finding increasing application. The challenge is greatest in the synthesis of significant quantities of human therapeutic proteins intended for clinical approaches that seek to correct deficiencies of endogenous biomolecules or to augment their existing levels. This approach requires high expression levels, stable production capacities, products whose safety and efficacy are rigorously controlled and products of precisely defined structure which accurately reproduce that of the naturally occurring proteins. Despite their unequalled potential to fulfil these criteria, it is only in recent years that animal cells have become useful as practical, industrial-scale bioreactors. This book charts the technical hurdles which have been overcome to achieve this aim, considers how the performance of animal cells can be optimized using present technology and considers some of the technical obstacles that still remain.

Some characteristics of bioreactors

A general scheme for the generation of products in a bioreactor is given in Figure 1.1. Overall, appropriate feed stocks or substrates must be

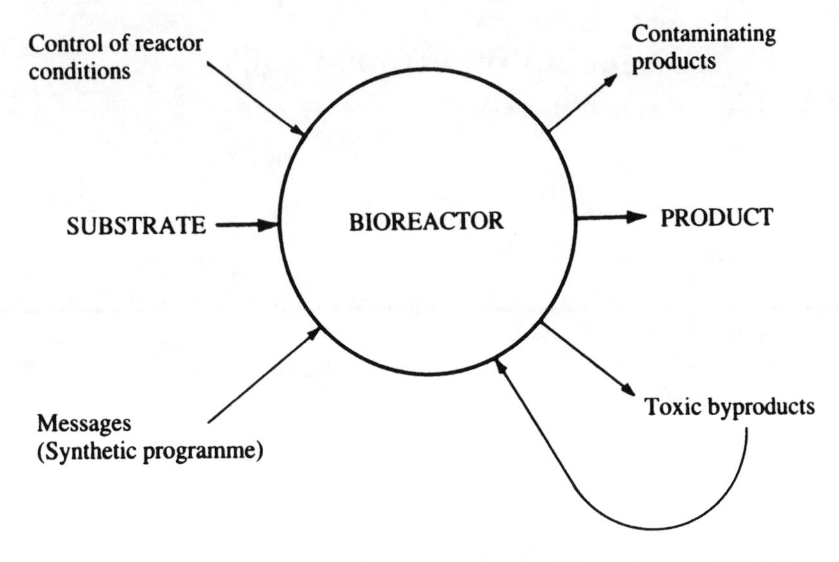

Figure 1.1 Generalized scheme illustrating the different factors which influence the performance of a bioreactor

fed to the bioreactor system in a controlled manner. After that the bioreactor must be able to maintain its own integrity during the course of the production run and must also synthesize the required product, which requires that the bioreactor contain the requisite information including signals and data which favour production of product. Effective control of the system is required to optimize yields and to regularize production so that product quality is maintained and downstream processing steps are served with an appropriate feed. The bioreactor may also produce by-products which contaminate the product stream and require elimination by purification. In addition, it may produce toxic products which limit its own efficiency and specificity. Production of these by-products must be reduced to a minimum, or an effective method for their continuous removal must be included. In some cases, bioreactor by-products may introduce issues of safety in addition to those of efficiency and then properly validated methods for their removal are obligatory.

These basic considerations can be transposed to consideration of the cultured animal cell as a bioreactor as indicated in Figure 1.2.In this scheme a suitable culture medium is fed to cells which contain the necessary synthetic information. For animal cells, this medium is invariably complex and may require undefined elements such as animal serum in addition to simple nutrients. Extensive medium development work may be required to arrive at a satisfactory formulation. After the bioreactor has become fully populated with cells, more medium, possibly

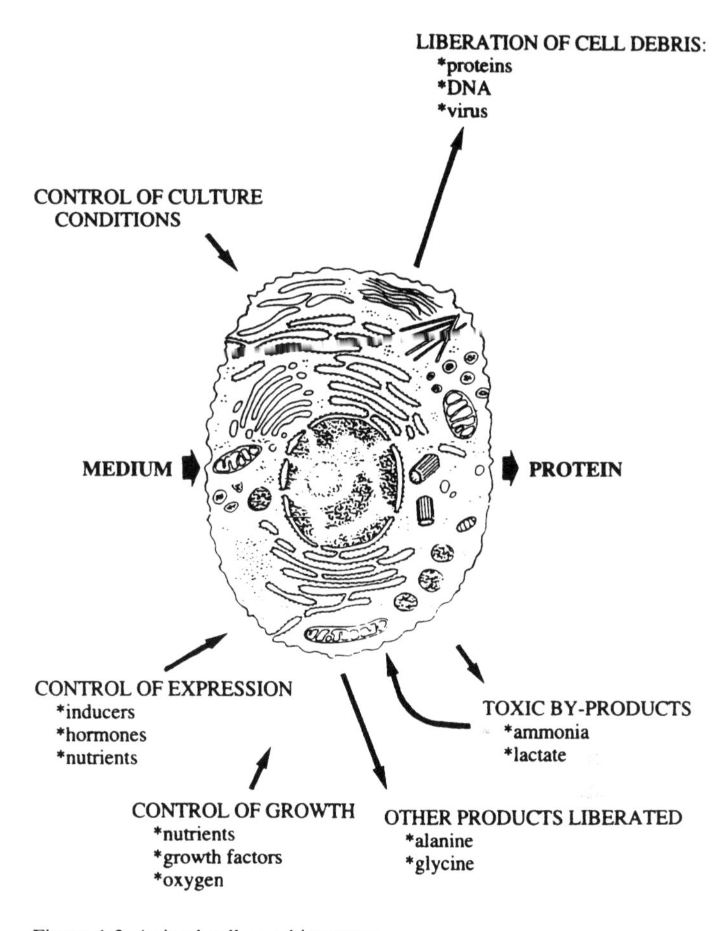

Figure 1.2 Animal cell as a bioreactor

of different composition, is then fed to the cells to maintain them in optimum condition for product generation. The scale at which the product is required is a determining factor for the type of fermentor configuration in which the cells are grown, the composition of the medium used and the way in which the information needed for synthesis is presented to the cells, that is, what expression system is used. Medium formulation and control of culture conditions are selected to limit the level of toxic products liberated, particularly ammonium ions and lactate, whose accumulation represents the main limitation in many animal cell cultures. Contaminating substances that can cause regulatory concern, including residual DNA, viruses and host-cell proteins also must be considered in the construction or choice of the animal cell to be used,

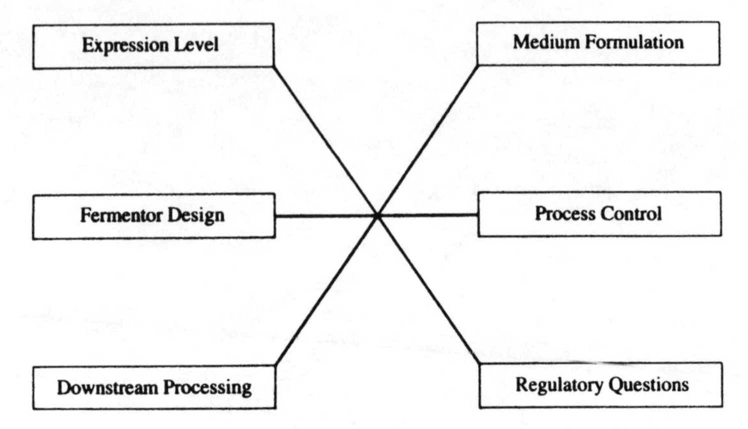

Figure 1.3 Interactive factors which need to be optimized for the efficient use of animal cells as bioreactors

the culture conditions employed and the downstream processing required after the production phase.

This book addresses these aspects of product generation in animal cells and considers how the cells' synthetic capacity can best be harnessed and controlled to give an efficient bioreactor. The parameters considered are summarized in Figure 1.3. A critical aspect of these parameters is their interactive nature: each parameter depends on the others to a greater or lesser extent, and the successful realization of an industrial process based on animal cells requires an integrated multi-disciplinary approach.

Experience with animal cells in the production of biologics

The use of animal cells as a production vehicle has had a turbulent history. The prehistoric phase (in terms of biotechnology) was the use of cellular substrates for the growth of viruses in the vaccine industry. This industry began with unsophisticated approaches, for instance using tissue fragments as sources of living cells which acted as a substrate for viral replication. Indeed, such methods are still in use today, as in the production of foot and mouth disease vaccine by the Fraenkel method, which uses epithelium from bovine tongue for growing virus. Need for enhanced reproducibility and regularity of production and for more sophisticated vaccines led to the development of cell culture for vaccine production. In the 1950s the urgent need for polio vaccine led directly to the use of primary monkey kidney cells for production of the Salk vaccine. Endogenous simian viruses (notably SV40) which contaminated these cells were subsequently shown to be present in vaccine preparations. The scare produced by this observation stimulated research into the use of exhaustively characterized human diploid cells for vaccine production, culminating in WI38, the prototypical human diploid fibroblast which has been used for the production of a wide range of vaccines and has remained the industry standard cell substrate for over 20 years.

The vaccine industry thus served as the prototype or pilot, both for the development of hardware for large-scale cell culture and, importantly, for the regulatory approach taken to culturing animal cells and the production methods associated with them. Vaccines still represent the largest use of animal cell culture systems, and the production of foot and mouth disease virus still represents the biggest single example of this technology in industrial use. In a single year over 2 million litres of BHK cells in culture were used for this product by one company alone (Wellcome) (Radlett, Pay, & Garland, 1985)

The rise of genetically engineered microbial systems and their limitations

Apart from vaccine production, cell culture was seen by many workers simply as a research tool which required difficult, complicated and expensive techniques due to problems of maintaining sterility, cell growth and the cost of the medium supplements required. Each cell had its own idiosyncrasies and each laboratory had its preferred medium and methods, making cell culture almost an art and certainly not a process adapted to industrial utilization. Indeed, in the late 1970s, when production of heterologous protein in bacteria was demonstrated, it was confidently predicted that cell-culture methods would eventually be completely superseded by recombinant DNA technology using microbial host systems. These would produce limitless, low-cost supplies of protein using well-understood fermentation techniques and would soon render the fastidious and expensive cell-culture approach obsolete.

Nowadays however, animal cells used as bioreactors for the production of recombinant proteins represent an increasingly important segment of the biotechnology industry, and systems using animal cell technology account for over half of the revenue generated by the new biotechnology (Table 1.1 and Spier, 1992). To what is this success due and how have animal cells staged a comeback against the microbial systems that appeared to be so much better adapted to industrial exploitation?

The main reason why bacterial and yeast systems have not proved satisfactory for the production of all proteins is their incapacity to reproduce mammalian proteins with complete fidelity. This is because, even after accurate translation of the relevant mRNA, many proteins produced in animal cells undergo a number of post-translational chem-

Table 1.1 *Currently licensed pharmaceuticals produced from cultured animal cells*

Product	Clinical Use	Production Cells Used	Production System Used
Alpha-interferon	Cancer, viral infection	Namalwa	Stirred Tank
Beta-interferon	Cancer, viral infection	CHO	Various
Erythropoietin	Anaemia	CHO	Roller bottles
Tissue plasminogen activator	Thrombolysis	CHO	Stirred Tank
Factor VIII	Haemophilia	CHO	Various
Hepatitis B Surface Antigen	Vaccine	CHO	?
Human Growth Hormone	Dwarfism	C127	?
Granulocyte colony stimulating factor	Chemotherapy rescue	CHO	Stirred Tank
Monoclonal Antibodies:			
Orthoclone OKT3	Kidney transplantation	Hybridoma cells	Stirred Tank
Centoxin*	Septic Shock	Hybridoma cells	Perfused suspension culture
Cell Line	CHO - Chinese Hamster Ovary Cells Namalwa - Namalwa Human Lymphoblastoid Cells C127 - Mouse Mammary Cells		

* Currently withdrawn in some countries due to lack of clinical efficacy

ical modifications or processing steps before the mature protein is secreted. These modifications cannot all be accurately reproduced by microbial systems.

A considerable range of these so-called post-translational modifications are encountered in different proteins. Glycosylation occurs very frequently, particularly in secreted proteins, but other modifications such as phosphorylation, sulphatation and carboxylation at specific residues, incorporation of lipid moieties or maturation by selective protein cleavage may also occur. The types of post-translational modifications most frequently encountered in animal cells are summarized in Table 1.2.

These modifications may be essential for the biological activity of many proteins, and, if not directly involved in activity, they can profoundly affect the stability of the protein in vivo, its clearance rate and route and its distribution in the tissues. These effects may appear paradoxical, for example, erythropoetin (EPO), the stimulator of red blood cell production, is completely inactive when produced in an unglyco-

Table 1.2 *The post-translational modifications most frequently encountered in mammalian proteins*

Post-Translational Modifications
* Glycosylation
* γ-carboxylation of glutamic acid
* β-hydroxylation of aspartic acid
* Phosphorylation and sulphatations
* Proteolytic processing
* Amidations

sylated form (Dube, Fisher and Powell, 1988). In contrast tissue plasminogen activator (tPA) suffers no activity loss if produced without its post-translational sugar moieties, and its biological efficacy is increased several fold because of the resultant prolonged circulatory half-life (Cambier et al, 1988). This arises because one of the normal clearance mechanisms for tPA involves its capture by hepatocytic carbohydrate receptors and its subsequent elimination from the circulation (Lucore et al, 1988). This phenomenon has resulted in the creation of several mutant forms of tPA, specifically designed to have augmented half-life, in which site-directed mutagenesis has been used to eliminate the offending glycosylation sites (Cambier et al, 1988). In immunoglobulins, glycosylation has little or no effect on the specificity and avidity of antigen binding but is essential for cytotoxic activities mediated through the Fc receptor, probably as a result of effects on the tertiary structure of the antibody. Bacteria are incapable of glycosylating proteins and, obviously therefore, can never produce authentic mammalian glycoproteins. On the other hand, yeast can glycosylate protein and can recognize and respond to the same glycosylation signals in an amino acid sequence as animal cells. However, the nature of the sugars added by yeast differs from those normally supplied by animal cells. Again, differences from normal antigenicity and pharmacokinetics can result in subsequent major impacts on effective biological activity and safety.

Glycosylation is an enzymic process and, as such, is governed by the enzymes available in the host cell and the availability of substrate. For this reason, as discussed in a later section, glycosylation can vary with conditions, even in one given cell type. Thus, reproducible glycosylation requires rigorous control of culture conditions, and synthesis and validation of correctly glycosylated proteins remain one of the major challenges of health-care biotechnology. Complications may also arise at other levels when heterologous proteins are expressed by microbial systems. In some cases, the protein itself may be toxic to the host cells, resulting in low levels of expression (although this problem can, in principle, be mitigated by the use of tightly controlled promoters which allow effectively no expression of the inserted gene until de-repression conditions are applied (Podhajska et al, 1985). A related problem may be uncontrollable proteolytic cleavage of the required product which can occur in *E. coli* or yeast. In some cases the induction process may result in co-expression of defensive proteases by the micro-organism (Kane and Hartley, 1988). In *E. coli* particularly, many proteins are produced as insoluble aggregates that, although sometimes produced at high level, present major difficulties for the recovery of the active product (Goeddel et al, 1979). Other problems are sometimes encountered which derive from differences in the translation control mechanism between bacterial and animal cells. Sequences in the structural gene for the mammalian protein are sometimes perceived as instruction signals by the micro-organism. In *E.coli*, a particular problem is the so-called "restart" phenomenon in which internal sequences in the structural gene behave as translational initiation signals which cause artifactual translation starts that interfere with correct translation and which can give rise to production of a significant proportion of the required proteins in a truncated form (Preibisch et al, 1988; Amann et al, 1988). This obviously lowers productivity, but its main impact may be in downstream processing in which the presence of variant forms of a protein may render adequate purification and analysis very difficult. This problem can be largely resolved by modifying the sequence of the mRNA to minimize recognition of such sequences, but the efficiency of translation may be reduced by such manipulations.

The restart phenomenon has been reported for the expression in *E. coli* of several proteins of potential clinical importance including IFN, TNFα, Factor XIII and p21 ras (Preibisch et al, 1988; Amann et al, 1988).

Acceptance of animal cell technology for industrial production

Thus, there are compelling reasons for using animal cells as bioreactors for making recombinant animal and human proteins. However, their acceptance as a commercially realistic production tool is a recent phenomenon. This is because of a series of difficulties, real and perceived, relating to the management of animal cells in production. These difficulties involve the following points (see also Table 1.3):

Table 1.3 *Perceived difficulties with the use of animal cells as bioreactors for product generation*

* Low specific yields of product (pg/cell/day)
* Medium expensive, variable and a serious potential source of contamination
* Fermentation equipment required (for high biomass) is complicated and costly
* The inherent fragility of the cells means that they can be damaged by shear forces during fermentation
* Process control difficult
* Not all cell types are acceptable to the regulatory authorities
* Purification may be difficult due to high levels of exogenous protein in the medium used

1. Animal cells generally produce lower quantities of recombinant protein per cell than the best microbial systems. The perceived limitations of animal cells in this respect and the approaches used to maximize their capacity for protein expression and secretion are discussed in Chapter 2.
2. Generation of the necessary high levels of biomass requires more complex fermentation and control equipment for animal cells than for microbial cultures. Due to the fastidious nature and slower growth of animal cells, they also require a higher standard of engineering skill and investment to produce plant capable of operating efficiently and without losses from microbial contamination.
3. The effective control of culture parameters is essential in animal cell culture if reproducible results are to be achieved. However, current knowledge of cell metabolism may be insufficient to achieve this, and appropriate and efficient sensors and monitoring systems are lacking. As discussed in the section on control of cell cultures (Chapter 3), these areas are currently under intense study, and new methods of modelling and monitoring

cell cultures are now being evaluated. An enhanced understanding of the ways in which cellular metabolism can be influenced in favour of greater product generation (reviewed in Chapter 4) will also improve the efficacy of animal cells as bioreactors.

4. The medium habitually used for animal cell growth may be poorly defined and contain contaminating protein in large quantities which makes efficient purification very difficult. In addition, the medium, and particularly the animal serum with which it is supplemented, may be contaminated with viruses, mycoplasma or other adventitious agents. Effective quality control to eliminate such contaminated raw materials is difficult and expensive. Batch to batch variability of serum can also profoundly and adversely affect the reproducibility of the production process. The design of effective, chemically defined culture media which avoid these technical difficulties is discussed Chapter 3.
5. Past reticence of the regulatory authorities to permit the use of certain abnormal cell types and the complexity of the tests prescribed for the characterization of cells have hindered development of some of the most efficient animal cell production systems. The current approaches to assuring the safety of biopharmaceuticals derived from animal cells and the techniques employed to achieve this are examined in Chapters 5 and 6.

Intense multidisciplinary efforts have been made and continue to be made to alter the biologics industry's perception of cell culture as a delicate, unstable and ill-understood system that is governed more by the art of the tissue culturist than by science. As a result, many aspects of cell culture, including the nutritional requirements of cells, bioengineering needs and regulatory issues have been placed on a rational footing which permits cost-effective industrial exploitation. Figure 1.3 summarizes the various interactive elements which must be optimized to achieve effective production in animal cells.

2 Yields of Recombinant Product: Engineering Cells for Maximum Expression

In the case of virus vaccine production, the actual mass of product required, even in the largest-scale application, is low because a little virus goes a long way. However, when the requirements for therapeutic proteins are considered, a different level of productivity becomes necessary because kilograms or tens of kilograms of pure protein may be needed. To avoid completely impractical culture volumes, the specific production rate of protein by the cells (i.e. picograms of protein per cell per hour) must be increased greatly over that which is normal for animal cells, which, with a few notable exceptions such as gamma globulin secreting cells, globin secreting cells and the cells responsible for the production of some digestive enzymes, do not produce protein at a very high rate.

Two basic approaches can be used to achieve high yields of secreted proteins from animal cells. The first involves tapping the existing potential of some cells to produce and secrete large quantities of a given protein under the control of pre-existing synthetic machinery. This is the situation which exists in antibody-producing cells in which high levels of antibody production are obtained due to the presence of powerful transcriptional enhancers for the immunoglobulin genes. The hybridoma technology proposed by Kohler and Milstein (1975) combines this capacity with the immortality of myeloma cell lines. This technique uses somatic hybridization of primary antibody-producing spleen cells with serially propagated myelomas to permit the long-term production of high levels (up to 500 μg/ml) of the required monoclonal antibody in hybrid cells that can be readily adapted to large-scale culture in fermentors.

Hybridomas produced in this way are selected by growing them in medium containing hypoxanthine, aminopterin and thymidine (HAT medium) which permits only the long-term growth of successfully fused cells which express hypoxanthine–guanine phosphoribosyl transferase (Kohler and Milstein, 1975).

Many hybridomas lose chromosomes soon after fusion, and successive cloning cycles are used to select productive hybridomas that appear to be stable. Loss of productivity in hybridomas can also occur much later when the cells are being used for production, still at least partially due to loss of chromosomes containing the antibody chain gene loci. One approach to limiting such chromosome loss has been the use of spleen cells from Robertsonian (8.12) 5Bnr mice whose heavy chain Ig locus is linked to a selectable marker (adenosine phosphoribosyl transferase) which allows selection pressure for maintenance of the antibody gene to be applied by growing the hybridomas in HAT medium (Taggart and Samloff, 1983). However, this approach is not practical in large-scale production systems, and periodic recloning and careful control of culture conditions to minimize the emergence of nonproductive cell populations are the methods more generally applied (Frame and Hu, 1990).

The second approach to high-yield animal cells requires the engineering of a recombinant cell in which expression of the gene of interest is driven artificially by placing it under the control of efficient promoter and enhancer signals. Much ingenuity has been devoted to the harnessing of the most powerful promoters and the most efficient vectors, and a wide variety of high-performance systems has been devised. Many of the most efficient transcription systems are based on viral regulatory sequences such as those from SV40 (Mulligan and Berg, 1981), retroviruses (Chang et al, 1980; Gorman et al, 1982), bovine papillomavirus (Lusky et al, 1983) and cytomegalovirus (Boshart et al, 1985). Although the regulatory elements from these different viruses show little sequence homology, they are active in a range of different cell types and have been widely used in consequence. Tissue specific enhancers such as the immunoglobulin heavy-chain enhancer (active in lymphoid cells, Neuberger, 1983) and the human beta globin locus control region (active in erythroid cells, Grosveld et al, 1987) have also proven useful for driving protein expression in animal cells. A full review is beyond the scope of this work but the main options are discussed next.

Transient expression systems

Many animal cells give short-term or transient expression of heterologous proteins when plasmid vectors are introduced into them. The exogenous DNA in such systems is associated with the host cell nucleus but is not integrated into the host cell chromatin. Transient expression systems are very useful as research tools when small amounts of protein are required for study or when the regulation of gene expression is being investigated. Several of these systems have been engineered to achieve very high copy numbers in the host cells, resulting in an intense burst of protein synthesis over a period of 24–48 hours after transfection.

SV40/COS

This system is based on COS monkey cells which are transformed with an origin-defective SV40 mutant and which express SV40 T antigens. By providing the T antigens, COS cells permit the replication of any vectors containing the SV40 origin of replication. Such vectors can then replicate extrachromosomally to very high copy numbers, 1–5 x 10^5 copies per cell (Rigby, 1982). Significant quantities of recombinant proteins can thus be produced. For example, Whittle et al (1987) expressed up to 500 ng/ml of a mouse–human chimaeric antibody over a 72-hour period using transient expression in COS cells.

Vaccinia

Vaccinia virus replicates in a wide range of cells in culture and has been used to obtain high-level transient expression of heterologous proteins (Moss and Flexner, 1987). A potentially important advantage of vaccinia is that the large size (180 kb) of the viral DNA permits the insertion of at least 25 kb of heterologous DNA by homologous recombination. This strategy has been applied to production of HbSAg and of influenza haemagglutinin (Smith, Mackett and Moss, 1983; Smith, Murphy and Moss, 1983). The system has been particularly applied to the in vivo expression of mutant vaccinia virus particles bearing specific antigens from pathogenic organisms which may have value as live-virus vaccines. In this connection, the large size of exogenous DNA insertion available is of particular interest because, in principle, this would enable several foreign proteins to be expressed simultaneously and therefore permit the production of polyvalent vaccines using a single virus particle (Moss and Flexner, 1987). Vaccinia virus vectors for gene expression have recently been reviewed by Smith (1991).

Retroviruses

A third approach to transient expression has been to use vectors containing the Rous sarcoma virus LTR in a variety of cell lines (Browne et al, 1988).

Baculovirus

As already stated, vectors based on lytic viruses have not proved practical for the production of large quantities of recombinant proteins because of the self-limiting nature of the system, the host cells being destroyed as the virus completes its replicative cycle. One exception to this is the baculovirus expression system used in insect cells.

Baculoviruses have a host range which is essentially limited to lepidopteran species. In the late stages of baculovirus infection, the nuclei of infected cells become packed with virus containing inclusion bodies called polyhedra. The function of the polyhedra is to protect the virus from the environment between cell lysis and encountering another sus-

Table 2.1 *Advantages and possible disadvantages of baculovirus expression in insect cells*

Advantages	Disadvantages
Rapid high expression level	Expression is late in the lytic cycle when the polyhedrin promoter is used (but alternative promoters for earlier genes are available)
Correct splicing	
Recognition of mammalian glycosylation signals	Different oligosaccharide trimming reactions give altered glycosylation patterns
Freedom of insect cells from potential human pathogens or oncogenes	Heavily glycosylated proteins may be retarded during ER transit
Ability to produce proteins which would be toxic in mammalian cells	Insect cells may secrete large amounts of protease causing product degradation
Expression of mammalian cell surface receptors without the complicating presence of endogenous receptor	

ceptible host. Protection is achieved by embedding virus particles in a crystalline matrix of a 29 kDa virus-encoded protein, polyhedrin, which is stable in the environment until it reaches the alkaline midgut of an insect host, when it dissolves and releases infectious virus. In late infection, 20% of the weight of the insect may be polyhedrin.

Baculovirus also replicates in susceptible insect cells in culture and in a suitable system, polyhedrin may account for 50% of the cell's total protein and reach concentrations in the culture approaching 0.5–1.0 g per litre.

Polyhedrin expression is controlled by the strong polyhedrin promoter which becomes activated only very late in the baculovirus replicative cycle, and whose only known function is to drive polyhedrin synthesis. Polyhedrin is not an essential protein for virus replication and the polyhedrin promoter was therefore rapidly identified as a possible means of driving the expression of high levels of heterologous protein in insect cells (Smith, Summers, and Fraser, 1983; Pennock, Shoemaker and Miller, 1984). The expression of the very late virus gene promoters occurs in cells after the maturation of budded, infectious virus particles. Consequently, if a cytotoxic protein is to be synthesized this can be achieved without affecting virus replication adversely.

Insect cell/insect virus systems offer several advantages for the production of proteins for clinical use as summarized in Table 2.1. A primary advantage is that insect cells are capable of most of the post-translational modifications performed by mammalian cells. However, N-glycosylation appears to be different in that the oligosaccharides of insect cell derived glycoproteins tend to terminate in mannose rather than in sialic acid as in mammalian cells (Kornfeld and Kornfeld, 1985; Kuroda et al, 1990).

Sissom and Ellis (1989) also observed some limitations in the capability of insect cells to perform the requisite post-translational modification of mammalian proteins. In this study, the extracellular domain of the human insulin receptor was expressed using a baculovirus vector. This protein requires correct glycosylation before it can be proteolytically cleaved into its alpha and beta subunits which subequently reassemble to form an active heterodimer. In the baculovirus system, poor proteolytic processing and slow intracellular transport was obtained leading to considerably reduced yields. Nevertheless, sufficient quantity of the mature protein to complete the study (several milligrams) was finally recovered (Sissom and Ellis, 1989).

Kuroda, Veit and Klenk (1991) have also reported that highly glycosylated proteins such as influenza virus haemagglutinin are secreted by insect cells with sugars of lower complexity than those in the authentic protein and that other processing events may be retarded.

As well as expressing inserted cDNA, the baculovirus/insect cell system is able to express intron-containing chromosomal genes correctly (Iatrou, Meidinger and Goldsmith, 1989).

Importantly, insect cells are generally perceived as being a 'safe' substrate for pharmaceutical production because they cannot be host to the viruses and the prion-type agents which are currently the focus of much concern when mammalian production cells are used.

Insect cells lend themselves to mass cultivation techniques although, as will be discussed in a later section, they may require culture conditions somewhat different from those normally applied to mammalian cells.

Other baculovirus promoters have been used besides the polyhedrin promoter. These include that for the p10 gene which encodes another abundantly expressed but nonessential baculovirus late protein and promoters for earlier proteins such as the AcMNPV 6.9K basic protein and the vp39 major capsid protein (Hill-Perkin and Possee, 1990).

Other refinements of the baculovirus system include the utilization of easily evaluated markers to allow selection of recombinants rather than the time-consuming identification of non-polyhedrin producing plaques by visual inspection. One approach to this has been the construction of vectors which permit co-expression of beta galactosidase with the required protein (Vialard et al, 1990).

Transient expression in insect cells has proved especially useful for the production of proteins such as activated oncogene products which would be toxic in mammalian cells. Expression of mammalian cell surface receptors in insect cells (where there are no complications due to the presence of endogenous receptors) has also proved very useful as a research tool.

Stable expression systems

Although transient expression systems can provide useful quantities of protein for research purposes, stability of expression is a critical requirement for any manufacturing process and can be achieved only by the development of stable transfected cell lines. This objective can be attained either by using stable virus vectors which replicate the required gene independently from the host cell genome or by stably integrating the cloned gene into the genome of the host cell. Both approaches have advantages and both exhibit problems which require skilled genetic engineering to overcome.

Stable extrachromasomal virus vectors

Most of the viral replicons which were initially used for the expression of exogenous genes in animal cells suffer from the central problem that gene amplification is accompanied by cell death as the virus completes its replicative cycle. An ideal virus vector would replicate continuously and be independent from host cell chromosomal control, thereby continually expressing the required protein product, but would not produce cell lysis.

Bovine papilloma virus
Bovine papilloma virus (BPV) is one virus which can fulfil these requirements. BPV is a small DNA tumour virus which induces epithelial and mesenchymal tumours in vivo and which transforms bovine and murine cells in culture. The BPV genome is a double-stranded circular DNA of 7.9 kb which has the unusual property of replicating as an extrachromosomal, multicopy plasmid in rodent cells. Replication of the entire BPV genome or of a subgenomic 5.4 kb fragment representing 69% of the genome (the 69% transforming or 69T fragment) causes morphological transformation in mouse cells. This property was initially used to select those cells which have received BPV DNA, but this approach restricted the usefulness of th system to those cells which exhibit focus formation. Subsequently a range of selectable markers have been introduced into BPV-1 vectors to widen the host range to include cells which do not express the transformed phenotype but in which BPV-1 still replicates episomally.

Selective markers employed in BPV-1 vector systems include the herpes simplex virus type 1 thymidine kinase (tk) gene (Lusky et al, 1983), the *E. coli* xanthine-guanine phosphoribosyltransferase (gpt) gene (Mulligan and Berg, 1981) and the Tn5 neomycin resistance gene (Law, Byrne and Howely 1983). The effects of these insertions are now briefly discussed.

Table 2.2 *Examples of stable expressions of heterologous protein using BPV-1 based vectors*

Eukaryotic gene expressed	Size of closed gene (kb)	BPV sequence in vector	Reference
Rat preproinsulin	1.62	69T	Sarver et al, 1981
Human β globin	7.6	69T	DiMaio et al, 1982
Rat somatotropin	5.8	69T	Kushner et al, 1982
Human somatotropin/ mouse metallothionen promoter	3.5	69T	Pavlakis and Homer, 1983
Human β interferon	1.6	69T	Zinn et al, 1982
Human β interferon	1.6	69T	Mitrani-Rosenbaum et al, 1983
MMTV LTR + H-MSV vras	3.7	69T	Ostrowski et al, 1983
HLA heavy chain + human β globin	8.5 + 7.6	69T	DiMaio et al, 1984
Hepatitis B surface antigen	2.7	Whole genome	Stenlund et al, 1983
Histone			Green et al, 1986
tPA			Bendig et al, 1987

Efficient BPV-1 shuttle vectors have been constructed which are propagated as extrachromosomal elements in both animal and bacterial cells (DiMaio, Treisman and Maniatis, 1982). Most BPV vectors have used the 69T BPV fragment, and this has permitted insertion of up to 16 kb of exogenous DNA (Table 2.2). In the transformed cell these vectors replicate as multicopy circular episomes present at 10–200 copies per cell (Campo and Spandidos, 1983). With many foreign proteins high-expression levels of the order of 10^6–10^8 molecules per cell per day have been reported (Stephens and Hentschel, 1987). All of the gene products described in Table 2.2 were accurately and efficiently transcribed and in most cases, the proteins were correctly processed and expressed.

Despite the widespread use of BPV-1 vectors for the production of recombinant proteins, several mechanistic aspects of vector function remain obscure and, in some cases, give rise to practical problems. Many studies have shown that rearrangements, integrations and deletions which affect primarily the bacterial sequences are frequent events in BPV-1 vectors. The BPV-1 vectors containing the human β-interferon gene described by Zinn et al (1982) and Mitrani-Rosenbaum et al (1983) both acquire exogenous DNA and exist in some cells as multimeric forms. Likewise the HBVsAg-containing plasmid described by Stenlund et al (1983) was also affected by both deletions and insertions.

Although, as mentioned earlier, the use of selectable markers can expand the host range that is usable with BPV-1 vectors by providing a means of selection other than the appearance of the transformed phenotype, use of such markers greatly alters the state of the vector in the host cell and favours integration of the plasmid DNA into the genome, leaving only a small proportion as episomes (Lusky et al, 1983; Mulligan and Berg, 1981; Sekiguchi et al, 1983, Bostock and Allshire, 1986).

However, BPV neomycin resistance plasmid vectors such as pd BPV-MM Tneo described by Law et al (1983) have proven particularly useful because in general, they escape this limitation and are maintained as unrearranged episomes in transformed cells. They are also able to shuttle between the eukaryotic cells and bacteria (Law et al, 1983; Matthias et al, 1983; Meneguzzi et al, 1984). For reasons which remain unclear, even neoresistance plasmids become predominantly integrated in a few cell types (Bostock and Allshire, 1986).

One factor which appears to govern the intracellular fate of BPV-1 vectors is the method used to introduce the DNA into the host cell. Microinjection directly into the nucleus is apparently less damaging to the DNA than calcium phosphate mediated transfection and is reported to favour the maintenance of BPV-1 vectors as autonomously replicating episomes (Bostock and Allshire, 1986).

Epstein–Barr virus

Epstein–Barr virus (EBV) is a large, herpes-type virus which can infect and transform human B lymphocytes. EBV has a large, 172 kb double-stranded DNA genome, elements of which have been used to create another class of episomal vectors (Yates et al, 1984).

A 1.8 kb fragment of EBV DNA, termed oriP, and expression of the Epstein–Barr nuclear antigen (EBNA-1) have been shown to be required for the maintenance of EBV-based episomal plasmids (Yates, Warren and Sugden, 1985). Circular DNAs containing oriP, the EBNA-1 gene and the bacterial hygromycin B phosphotransferase gene as a dominant selectable marker replicate autonomously and without inte-

Table 2.3 *Selectable resistance markers for use with animal cells*

Selectable Gene	Selective Agent	Reference
HSV thymidine Kinase (tn)	Used in TK^-cells (e.g. L cells)	Mantei et al, 1979
E. coli xanthine-guanine phosphotransferase (gpt)	Mycophenolic acid	Mulligan and Berg, 1981
Tn5 aminoglycoside phosphotransferase (neo^R)	G418 (aminoglycoside antibiotic)	Colbere-Gerapin et al, 1981
E. coli hygromycin B phosphotransferase (hph)	Hygromycin B (aminocyclitol antibiotic)	Blochlingler and Diggelmann, 1984
E. coli tryptophan biosynthesis (trpB)	Tryptophan deficiency	Hartman and Mulligan, 1988
Salmonella histidine biosynthesis (tn)	Histidine deficiency	Hartman and Mulligan, 1988

gration in several dog, monkey and human cell lines, although the plasmids were not retained in several rodent cell lines. Thus, the EBV-derived plasmids exhibit a wider host range than BPV-based vectors.

Another potential advantage for EBV vectors is their capacity to infect lymphoblasts and other lymphoid cells which can easily be propagated in suspension culture and so lend themselves to mass-production systems. One limitation of the use of these vectors may be the lack of flexibility in the transfection methods and selectable markers that have been successfully employed (Sugden, Marsh and Yates, 1985).

Integrated stable expression systems

Stable expression of a heterologous protein can also be achieved when the required gene is integrated into the host cell DNA in a way which permits its continued expression. However, integration is an event which occurs with low frequency after transfection and the use of selectable markers permits selection of clones which contain the gene of interest. Some of the selectable markers employed with animal cells are summarized in Table 2.3. A detailed review of appropriate selectable markers and their use has recently been published by Gorman (1986).

Position effects

With most vectors, integration into the host genome apparently occurs at random sites. Different clones exhibit wide variations in the level at which they can express the required gene. This variation appears to be largely independent of the number of copies of the gene that may be present. Such variations are clearly imposed by the host cell and are probably related to the position in the host genome at which integration has occurred. Ninety percent of the animal cell genome exists as transcriptionally silent 'repetitive sequences', so there is a high probability that a given random insertion site will be 'lost' in such sequences. Elsewhere, control DNA elements flanking the inserted gene and other aspects of its environment including the local state of the chromatin greatly influence the level of gene expression.

Thus, unexpected results were obtained when myeloma cells, which constitutively express Ig mRNA as 6–9% of their total cytoplasmic poly A^+ mRNA, were transfected with vectors which used the Ig enhancer and promoter elements to drive the expression of the tPA structural gene or of exogenous Ig genes. The transfected tPA or Ig genes were both expressed with similar efficiency in this system, but in both cases expression levels were much below that of the myeloma's endogenous Ig gene. This suggested that the exogenous Ig regulatory signals were being used suboptimally (Weidle and Buckel, 1987).

When SV40 and immunoglobulin regulatory element-driven expression of tPA were compared in myeloma cells, the SV40 system was superior by a factor of 10–30, again indicating that the transfected Ig transcriptional signals were not being efficiently used (Hendricks, Banker and McLaughlan, 1988). Although direct comparison between the two systems is difficult in this study because the construction of the 3' and 5' untranslated regions were not identical, the general inference was that integration at specific sites in the genome is necessary to achieve expression levels comparable with that of the endogenous Ig genes (Hendricks et al, 1988).

High-yield expression in myeloma cells of several proteins, including immunoglobulins, lymphocyte surface antigens and interleukins, has been reported using vectors which include a modified Ig heavy-chain enhancer and an Ig kappa variable-region promoter (Figure 2.1). Using this system, product yields in excess of 100 μg/ml were obtained (Traunecker, Okiveri and Karjalainen, 1991).

Stably transfected insect cell lines have also been established which use baculovirus intermediate early gene promoters or the inducible *Drosophila* metallothionein promoter (Culip et al, 1991). Cell lines generated in this way have been shown to consistently produce tens of milligrams per litre of correctly processed mammalian proteins.

Amplification of integrated systems

Even when foreign genes packaged in the most efficient transcription units, including powerful promoter and enhancer sequences, are integrated into productive regions of the genome of the chosen host, they

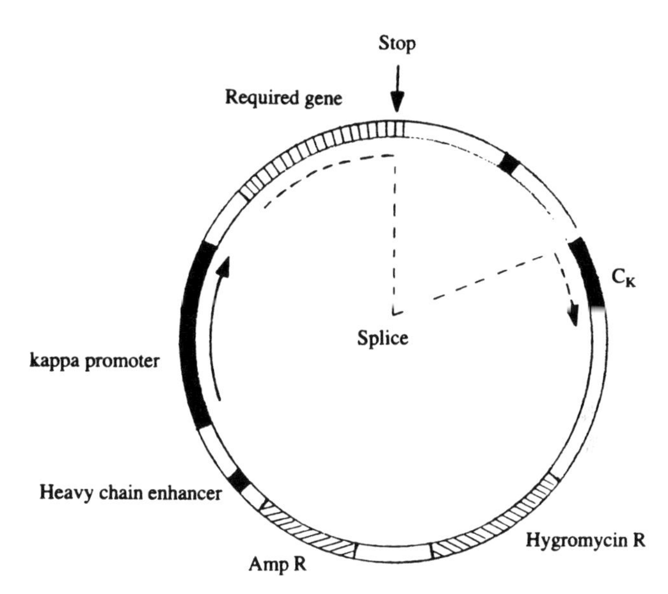

Figure 2.1 Schematic of expression plasmid for use in myeloma cells. The required gene is spliced into the genomic Ig_K control region gene (C_K) which also supplies polyadenylation signals. (After Traunecker et al, 1991)

are still rarely capable of producing useful quantities of protein. This is partly because the number of copies of the vector in the host genome typically remains low. Higher levels of expression can be achieved if the number of copies can be effectively increased. As we have seen, high copy number is achieved rapidly with episomal systems such as BPV- and EBV-based vectors, but these systems are limited both by the host range in which the virus can replicate and by the nature of the DNA sequence that can be stably maintained and expressed. Increasing the copy number of integrated genes should overcome these restrictions and can be achieved by the process of gene amplification.

It has been shown that one mechanism by which animal cells acquire resistance to certain toxic substances is by increasing the copy number of (or amplifying) genes whose products tend to nullify the effects of the toxic agent.

Dihydrofolate reductase

The prototype for this type of amplification has been the use of the antimitotic agent methotrexate (MTX) which functions by inhibition of dihydrofolate reductase (DHFR). Resistance to MTX is achieved by amplification of the DHFR structural gene (Schimke, 1978). Use of increasing concentrations of MTX can result in amplification of up to 1000-fold of the DHFR gene and, importantly, of segments of DNA exceeding 1000 bp adjacent to the DHFR gene which are usually amplified at the same time. Thus, great amplification of the required gene can be obtained when expression vectors are used which place the gene of interest in the correct context for co-amplification with DHFR (Schimke, 1978; Stark and Wahl, 1984) and when selective pressure is applied.

Vectors designed for amplifiable expression contain the following essential features:

1. Bacterial sequences permitting propagation and selection in bacteria
2. A selectable marker which functions in mammalian cells
3. An amplifiable gene
4. An expression cassette for insertion of the required gene

When possible, the amplifiable element should also serve as the selectable marker because the rearrangements and mutations which are liable to occur in transfected DNA could otherwise uncouple the selectable marker and the gene to be amplified. Another important factor is that effective amplification must be achievable at practical levels of the enzyme inhibitor. Limitations on this concentration can be imposed by the solubility of the inhibitor, by its cost or by its other toxic effects on the cells. Figure 2.2 shows a typical amplifiable vector designed for use with the DHFR/MTX system.

Using such vectors the level of recombinant proteins expressed broadly parallels the degree of amplification obtained in the system, but this relationship is by no means general and other factors may ultimately limit the level of protein production.

It is particularly useful to be able to perform amplification procedures in cell lines which lack the amplifiable gene. If endogenous enzyme activity is present, higher levels of inhibitor are required to achieve effective amplification. There is also the possibility that amplification of the endogenous enzyme gene may occur, thus reducing the effective amplification of the required gene. DHFR$^-$ mutants of CHO cells are available (Urlaub and Chasin, 1980), and these have been used in most systems employing DHFR amplification. DHFR$^-$ mutants of other cell types have been difficult to produce, and this has limited exploitation of the DHFR amplification system in other cells which might be better adapted than the CHO cell for some applications. However, it has been shown that effective amplification can sometimes be practically achieved even in cells which retain an endogenous functional DHFR gene (Okamoto et al, 1990).

Use has also been made of a mutant DHFR gene whose product is relatively resistant to MTX and which can therefore be selected for in

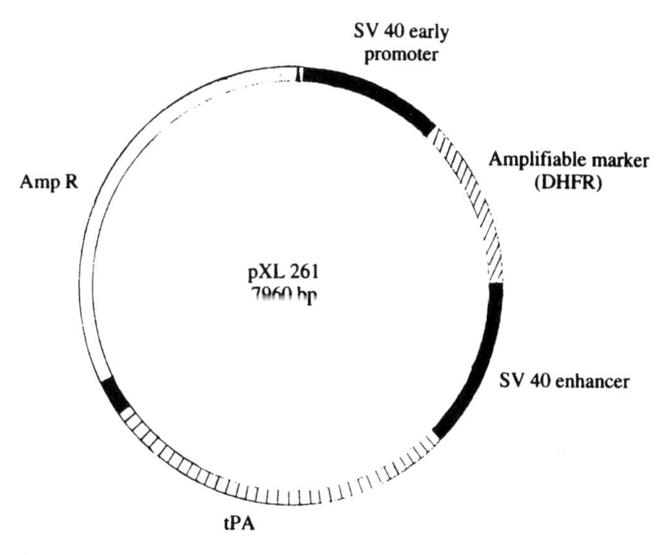

Figure 2.2 Schematic drawing of plasmid pXL 261 constructed for expression of tPA in CHO cells using DHFR as an amplifiable marker. The black segments in the diagram are SV40 sequences including the SV40 early promoter, enhancer and polyadenylation sequence.

wild-type cells which retain their endogenous DHFR genes (Haber and Schimke, 1982; Hendricks, Luchette and Barker, 1989).

In addition to its use in CHO cells this latter system has also been applied to heterologous genes expressed in myeloma cells under the control of immunoglobulin regulatory elements (Hendricks, Luchette and Barker, 1989).

Adenosine deaminase

Another amplifiable system involves the use of adenosine deaminase (ADA) which catalyses the conversion by deamination of adenine nucleotides to their inosine analogues. This enzyme is specifically inhibited by deoxycoformycin (dcf). Although ADA is not normally an essential enzyme it can be made to become essential if cells are cultured in the presence of a toxic excess of adenine or the toxic adenine analogue, 9-D-xylofuranosyl adenine. In these cases, ADA detoxifies the adenine nucleotides by deamination to their less toxic inosine derivatives. Use of these two approaches combined (dcf plus adenine excess) has been reported to amplify the vectors used to about 500 copies per cell without consistent increases in the copy number of the endogenous ADA genes (Kaufman et al, 1986).

Metallothionein

Metallothioneins are proteins which play a role in heavy-metal detoxification. Metallothionein genes can be amplified by use of selective medium containing cadmium (Hamer and Walling, 1982). The metallothionein genes themselves cannot be used effectively as a selectable marker in some cell types and the usefulness of this system is limited accordingly (Bebbington and Hentschel, 1987).

Glutamine synthetase
Another amplifiable selectable marker which makes use of the glutamine synthetase (GS) gene has been developed by workers at Celltech (Bebbington and Hentschel, 1987). Glutamine is an essential metabolite which plays a key role as an energy source and in the biosynthesis of macromolecules in animal cells in culture. The only pathway for glutamine synthesis is from ammonia and glutamate catalyzed by GS. Thus, in the absence of glutamine, GS is an essential enzyme and treatment of cells with methionine sulphoxime (MSX), a specific inhibitor of GS, is lethal.

Although, as discussed later, some cell types exist which are GS deficient and do not grow at all in the absence of exogenous glutamine, most types of cell do possess endogenous GS activity. Furthermore, GS levels in these cells may be up-regulated by glutamine deficiency so that these cells become adapted to growth on glutamine-free medium (Feng, Shiber and Max, 1990). However, it appears that the endogenous GS gene is not usually effectively amplified by treatment with MSX (Bebbington and Hentschel, 1987). This confers on the GS amplification the great advantage that high-level expression can be achieved without the need for the development of GS^- mutant cell lines.

This system has been used in CHO cells to obtain high-level expression of tissue inhibitor of metalloproteinase (TIMP) (Cockett, Bebbington and Yarranton, 1990). After initial experiments in a transient expression system to determine the optimum promoter for TIMP expression, plasmids were constructed using the chosen promoter (an enhancer promoter fragment from human cytomegalovirus) to create the most efficient TIMP transcription unit, and amplifiable selectable markers were inserted into this (Figure 2.3). The authors compared the efficiency of DHFR and GS amplifiable markers in these experiments. When the resulting selected clones were tested for TIMP production before amplification, GS clones showed much better expression than DHFR clones. After a single amplification step the best GS clone tested increased copy number of the TIMP gene by a factor of 30 with a concomitant increase of 12-fold in TIMP expression. TIMP levels remained much higher than the best achieved by MTX amplification. Further amplification of the TIMP gene did not, however, give higher TIMP yields, suggesting that some other factor governing TIMP expression had become saturated. In general the correlation between copy number and expression level of the protein was poor.

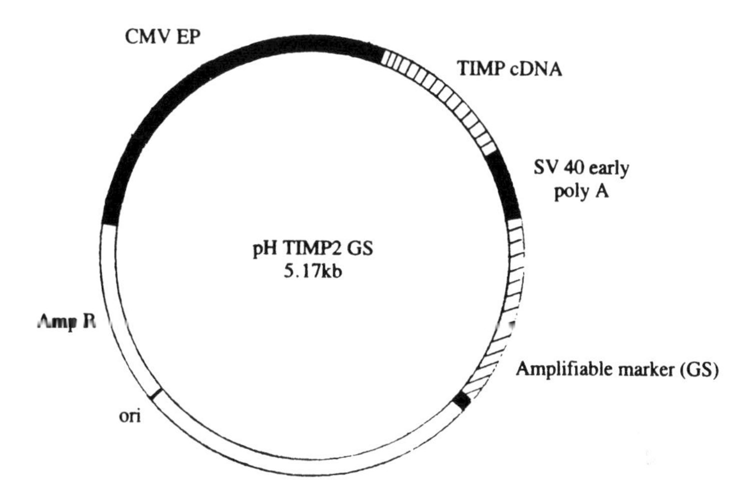

Figure 2.3 Schematic of expression plasmid using glutamine synthetase (GS) as an amplifiable marker (after Cockett et al, 1990). The pH TIMP2 plasmid containing the 5′ untranslated sequence and the immediate early promoter from human CMV (CMV.EP) was shown to give optimal expression of tissue inhibitor of metalloprotease (TIMP). For amplification experiments either the DHFR gene or the GS gene was incorporated as an amplifiable marker in the position indicated to give pH.TIMP2.DHFR or pH.TIMP2.GS, respectively (see text).

The same authors report high-level production (of the order of 100–200 mg/l) of several other recombinant proteins from CHO cells, including a human–mouse chimaeric antibody, fragments of this antibody and soluble CD4 (Cockett et al, 1990).

In a subsequent study (Bebbington et al, 1992) the successful use of the GS amplification system in NSO myeloma cells was reported. As previously discussed, vector amplification in myeloma cells using the DHFR system (Okamoto et al, 1990) has been achieved in some cases, but in the presence of endogenous DHFR activity, very high levels of MTX were required to obtain satisfactory amplification. In other studies, attempts to amplify antibody genes in myeloma cells using MTX did not produce significant amplification (Gillies et al, 1989). This is unfortunate because myeloma cells are a particularly desirable substrate for recombinant immunoglobulin production because they readily produce high biomass as suspension cultures in fermentors.

However, unlike the CHO cells previously discussed, myeloma and hybridoma cells appear to have an absolute requirement for glutamine due to their very low levels of endogenous GS activity. This is confirmed by the observation that they can be rendered independent of glutamine in the medium by transfection with a plasmid which produces GS expression (Bebbington et al, 1992). This suggested that amplification of GS containing selectable markers could be achieved at lower MSX concentration than would be necessary in cells such as CHO-K1, BHK21 or L cells which contain significant endogenous GS activity.

Expression vectors for immunoglobulin production using the h CMV-MIE promoter and the GS gene as an amplifiable selectable marker were incorporated into the NSO myeloma cell line which was chosen because of its low frequency of generation of spontaneous glutamine-independent variants. High levels of antibody production were achieved after a single round of MSX amplification, and production was stable for at least 65 generations in the absence of continued MSX selection. In fed-batch air-lift fermentors, antibody levels reached 560 mg/l, a level comparable to the yields of antibody obtained from hybridomas in industrial production processes (Bebbington et al, 1992).

These studies and other work published by the same group (Hassell et al, 1992) illustrate the point, made in Chapter 1, that integrated process development is essential for obtaining optimal results from cells as bioreactors. Thus, optimal expression was obtained using a defined glutamine-free medium which had to be supplemented with additional glutamate, asparagine and nucleotides to compensate for the absence of glutamine. The asparagine requirement of NSO cells is unusual and possibly was needed as a metabolic source of ammonia. The development of a suitable amino acid fed-batch regime also increased yield significantly and air-lift fermentation was shown to be superior to fermentation in stirred tanks. In other studies it has also been reported that methionine supplementation of the medium may be useful because MSX is partially detoxified through the methionine catabolic pathway and excess methionine can saturate this pathway, with a consequent sparing effect on MSX and thus effectively enhance its toxicity (Bebbington and Hentschel, 1987).

Tandem amplification

The availability of several gene-amplification strategies provides the possibility of independently amplifying different genes on separate vectors in the same cell. This approach may offer advantages in the production of proteins whose active form is composed of several subunits.

Tandem amplification has been used to generate antibody-producing CHO cells in which the required heavy-chain gene was co-amplified with the ADA gene using deoxycoformycin, and in a different population of cells, the light-chain genes were amplified using the DHFR/MTX system. Hybrid cells were then produced by fusion of a high light-chain producer clone with a high heavy-chain producer. The highest antibody producing clones (67 $\mu g/10^6$ cells/48 h) were then obtained from the hybrids produced by selection for increased resistance to both MTX and deoxycoformycin (Wood et al, 1990).

Tandem amplification has also been used in situations in which cooperation or interaction between different proteins is necessary to optimize yield. One example of this approach has been the co-expression of Factor VIII and von Willebrand factor (vWf) in CHO cells. When the Factor VIII gene was co-amplified with DHFR and expressed in CHO cells, yields of active protein recovered were much lower than those suggested by the amplification level achieved (Kaufman et al, 1989). A major reason for the poor recovery was the absence, in the serum-free medium used, of vWf, a serum protein which interacts with Factor VIII, resulting in the formation of a conformationally stabilizing complex and the protection of Factor VIII from proteolytic degradation.

The problem was overcome by taking cells which had previously been transfected with a DHFR–Factor VIII expression vector and in which the Factor VIII gene had been amplified by MTX treatment. A second expression vector which linked vWf and ADA was then introduced into these cells, and the vWf gene was co-amplified by treatment with deoxycoformycin. The level of recovery of active Factor VIII from the serum-free culture medium was then shown to increase corresponding to the observed increase in vWf expression (Kaufman et al, 1989).

Limitations of co-amplification systems

Expression vectors based on viral enhancer/promoter sequences and co-amplification of the required structural gene using drug-resistance markers as previously indicated have been widely and successfully used in research and industry. High-level expression of many useful proteins have been achieved, but despite these successes several factors persist which significantly limit the convenience and the practicality of amplification systems. The following approaches have been evaluated for their capacity to improve yields.

Favouring amplification events

One limitation of the standard systems is the time required to achieve effective amplification. Gene amplification occurs at a specific locus at a frequency of 10^4–10^6. To obtain useful amplification of the expression of the required protein it is frequently necessary to perform several cycles of cell culture in increasing concentrations of the selective agent to allow drug resistance by gene amplification to develop and resistant clones to be isolated. This process is time consuming and usually means that several months elapse between the initial transfection and the production of cells that are resistant to high concentrations of drug (Bebbington and Hentschel, 1987). For these reasons there has been considerable effort to identify factors which can increase the frequency of gene amplification.

Two groups have recently reported the existence of cis-acting elements in the mammalian genome which facilitate amplification when present on DHFR-expression plasmids introduced into animal cells. McArthur and Stanners (1991) identified a repetitive element termed HSAG-1 which increased the frequency of formation of MTX-resistant clones of CHO and HeLa cells when included in pSV^2 DHFR-based plasmids. The HSAG-1 element's amplification-promoting activity was localized to a 1.45 kb fragment containing Alu-like elements, inverted repeats and A/T rich regions (Beitel, McArthur and Stanners, 1991). HSAG-1 appears to provide both an increased probability of generating variants with increased copy number and a faster amplification cycle. It is not known how the HSAG-1 element functions, but McArthur and Stanners (1991) suggest that it may help to target the vector to recombination 'hot spots' in the genome.

The second amplification-facilitating factor reported is a 370 bp element identified in mouse ribosomal RNA which was shown to operate through interaction with a nuclear protein termed HMG-I (Zastrow et al, 1989; Wegner et al, 1989). Amplification-promoting activity in this case was localized to an A/T-rich region at the 5' end and an 11 bp palindrome at the 3' end. Again both the frequency of generation of transformants and the speed of amplification are increased. In this case too, details of the mechanism of activities remain unknown, although Zastrow et al (1989) favour an explanation based on a 'rolling circle' type replication event before integration into the genome.

A different approach to improving the rate of integration of incoming DNA into the receiving-cell genome has been reported by Wurm et al (1992). Their approach is based on the observation that, in common with many other rodent cells, CHO cells contain 500–1000 and 100–300 copies per haploid genome of sequences resembling A-type and C-type retrovirus genomes, respectively. Some of these sequences have been shown to be transcriptionally active since 0.2–0.5% of mRNA molecules in CHO cells are of retroviral origin (Anderson et al, 1990; Anderson et al, 1991).

Wurm et al (1992) tested the effects of introducing DHFR expression vectors containing genomic retroviral DNA sequences from CHO cells into $DHFR^-$ CHO cells transfected with separate DHFR plasmids constructed for MTX-amplified production of CD4-IgG (Byrn et al, 1989). The inclusion of a 2.9 kb A-type retroviral DNA sequence increased (by 5–20-fold) the number of MTX-resistant clones obtained in selective medium containing 200, 500 or 1000 nM MTX. A 6.6 kb C-type retrovirus sequence produced a more modest 3–4-fold increase at 500 and

1000 nM MTX. Shorter subfragments of these two retroviral sequences had a lesser stimulatory effect.

The effect of the 2.9 kb A-type retroviral sequence on expression of CD4-IgG and of tPA was also evaluated. In both cases, more producer clones were generated when the retroviral sequence was present, and the clones obtained expressed higher levels of the recombinant proteins. The increase in expression obtained was not affected by the orientation of the A-type sequence.

A possible explanation for the increase in productive clones might be that, despite their being introduced into the cells on separate plasmids, the gene of interest becomes associated with the retroviral sequence when plasmid molecules are ligated in the nucleus to form multimeric units before integration occurs (Finn et al, 1989). In these circumstances, integration into the genome, and, perhaps particularly, integration into transcriptionally active areas could be favoured by homology between the incoming retroviral sequences and multiple copies of these sequences in the genome (Wurm et al, 1992).

Stability of amplified sequences

A significant problem with amplifiable expression systems is that of the long-term stability of the amplified sequences. Amplified sequences are inherently unstable, although the degree of instability depends to some extent on the cell type used and the state of the sequences within the cell. In some cell types, amplified vector sequences may persist predominantly as small extrachromosomal genetic elements known as 'double minutes'. Because the double minutes possess no centromere, they are unevenly distributed between daughter cells at mitosis. Possession of lower numbers of double minutes confers a growth advantage on the cells so these amplified segments are rapidly eliminated from the cell population when selective pressure is removed.

More stable amplification occurs when the amplified sequences become integrated into chromosomes, and this is the situation which tends to predominate in the cell types most frequently used for amplified expression, such as CHO cells. However, high levels of amplification are almost invariably accompanied by significant chromosome abnormalities, and re-arrangements and deletions may continue to occur in amplified cells (Federspiel et al, 1984; Looney and Hamlin, 1987).

Production of stable, high-expressing cells thus requires careful selection and a considerable amount of time. Even then, loss of productivity over time may still occur even when selection pressure on the cells is maintained (Federspiel et al, 1984).

Recently, Cossons et al (1991) reported progressive loss of gamma interferon production in amplified CHO cells maintained in the continuous presence of 100 nM methotrexate. Interferon titre fell by over 70% over 120 days in culture. During this time the level of DHFR DNA fell markedly, although the level of interferon DNA fell rather less. No major chromosomal changes were observed. In parallel with this loss of productivity, maximum cell density obtained increased and cell viability was maintained for longer periods, suggesting that a growth advantage was operating for cells which had partially lost the amplified sequences.

Similar results were obtained by Raper et al (1992). In this case the protein being expressed in CHO cells was a humanized monoclonal antibody, Campath 1-H. Despite the continual presence of 100 nM methotrexate, a gradual decline in antibody yield was observed over a 200-day period. In this case also, some decrease in copy number of the amplified gene was observed.

These observations of loss of amplified sequences under continuous selection conditions were unexpected because previous studies (Wiedle et al, 1988) have indicated that such events are unlikely under continuing drug selection in the absence of major chromosomal alterations.

A possible explanation for this phenomenon is that cells may be able to develop drug resistance by other means than by amplification of the target enzyme. Other cellular metabolic responses which would confer resistance include the production of variant DHFR molecules less sensitive to inhibition by MTX (Haber and Schimke, 1982) or the selection of MTX transport mutants which would lower the intracellular concentration of MTX and so increase effective cellular resistance.

Thus, the mechanism of the cell's response to metabolic inhibitors may be more complex than was initially supposed and a combination of treatments aimed at several different aspects of the cellular response may be needed before truly stable amplified expression of heterologous proteins can be routinely achieved.

Dominant control regions

As we have seen, endogenous control elements which regulate the expression of eucaryotic genes often give higher levels of expression than the same element when re-introduced artificially into cells to drive expression of a required protein. This is frequently due to the 'position effect', in which expression levels are greatly influenced by the site of insertion into the genome. Yields can be raised by amplification, but this is time consuming, and the raised expression levels obtained may be unstable even when drug selection is maintained. However, some regulatory elements exist which can overcome or avoid such difficulties, and these have recently been employed to obtain very high-level expression of heterologous proteins.

Human beta globin expression is controlled by one such element. The region immediately flanking the globin gene comprises a 6.5 kb cluster of regulatory elements, including a promoter and two enhancers. These interact with trans-acting factors found in cells of the erythroid lineage and act alone or in combination to stimulate erythroid specific expression (Grosveld et al, 1987). The whole cluster is controlled by an element at the 5' end of the control region (Blom van Assendelft et al, 1989; Talbot et al, 1989).

Heterologous genes can be introduced into the human beta globin minilocus, and powerful expression can be obtained which is tissue specific, copy-number dependent, and *position independent*. Regulatory clusters with these characteristics are known as 'dominant control regions' (DCR).

In the case of the beta globin DCR, if the 5' element just mentioned is deleted, expression is greatly diminished and becomes position dependent. When used in MEL-C88 erythroleukaemia cells, induction of differentiation into erythroid cells by DMSO results in a 100-fold increase in heterologous gene expression (Grosveld et al, 1987).

This system has recently been used for the production of human growth hormone in MEL cells (McLean et al, 1992). Growth hormone production was maximal four days after induction with DMSO, and accumulated yields of 100 mg/l were obtained with the most productive clones. The human growth hormone was correctly processed and secreted. High-copy number of the transfected minilocus was obtained, and this remained stable for over 40 generations, even in the absence of any selection pressure.

A similar dominant control region has been reported to mediate T-cell-specific expression of the 'erythrocyte rosette receptor' (CD2) in transgenic mice. (Greaves et al, 1989). This 3' flanking control sequence again drives high-level expression in a position-independent manner.

Stimulation of expression by the use of vectors which incorporate dominant control regions clearly offers many advantages in terms of yield of protein, frequency of achieving productive clones and stability of production. It is very likely that other such regions associated with tissue-specific developmental expression of mammalian genes will be identified and that systems using these will become the method of choice for high-level expression of heterologous protein in animal cells.

Other factors governing heterologous gene expression

As we have seen, the first approach to engineering cells for maximum expression has been aimed at increasing the number of transcriptionally active copies of the required gene in the producing cell. Considerable efforts have been directed towards obtaining the gene of interest in a stably integrated form, in a genomic location favourable to high-level transcription and driven by efficient promoters and enhancers. Major successes have been achieved, and practical industrial expression of heterologous proteins in animal cells is now a reality.

In most of the systems discussed, transcriptional control relies on endogenous cellular transactivators to stimulate transcription from the strong promoters chosen to drive expression of the required hereologous gene. A system has recently been described by Hippenmeyer and Highkin (1993) in which transcription is increased by engineering cells to express a viral transactivator able to increase transcription from appropriate viral promoters.

BHK 21 cells were stably transfected with the herpesvirus transactivator, VP16. When the cells were then transfected with the gene for a heterologous protein under the control of a herpesvirus early promoter, greatly enhanced yields of heterologous protein were obtained relative to those obtained from equivalent constructions which did not express VP16.

Reported advantages of this system are that yields of several important recombinant proteins (including tPA) are at least as high as those achieved in gene amplified systems, and the time required to establish high-yielding cell lines is much less than that needed in amplification systems. Expression was reported to be stable over several months.

However, as might be expected, improved transcriptional activity cannot provide unlimited improvements in protein yield. Procedures which increase the copy number of the gene of interest eventually reach a plateau beyond which no increases in protein output can be achieved. Furthermore, maximum copy number may not correlate with maximum protein yield because it may be deleterious for the producer cell in other respects. Such effects are observed in microbial expression systems in which competition by the cloned gene for energy and for the limited intracellular pool of transcription factors are seen as being the main limiting factors (Seo and Bailey, 1985).

Several other factors also certainly play a critical role in animal cells. In particular, limitations on the expression of recombinant proteins may reflect the cell's inability to handle the necessary post-transcriptional events sufficiently rapidly or precisely to keep up with the increased supply of mRNA.

Following transcription, the mRNA is translated in the rough endoplasmic reticulum via complexes involving membrane-bound polyribosomes and soluble translation factors, secreted into the lumen of the endoplasmic reticulum (ER), glycosylated if appropriate and transported to the Golgi apparatus, further glycosylated, proteolytically processed in some cases and secreted (Figure 2.4). Any of these translational and post-translational events can become over-driven by greatly en-

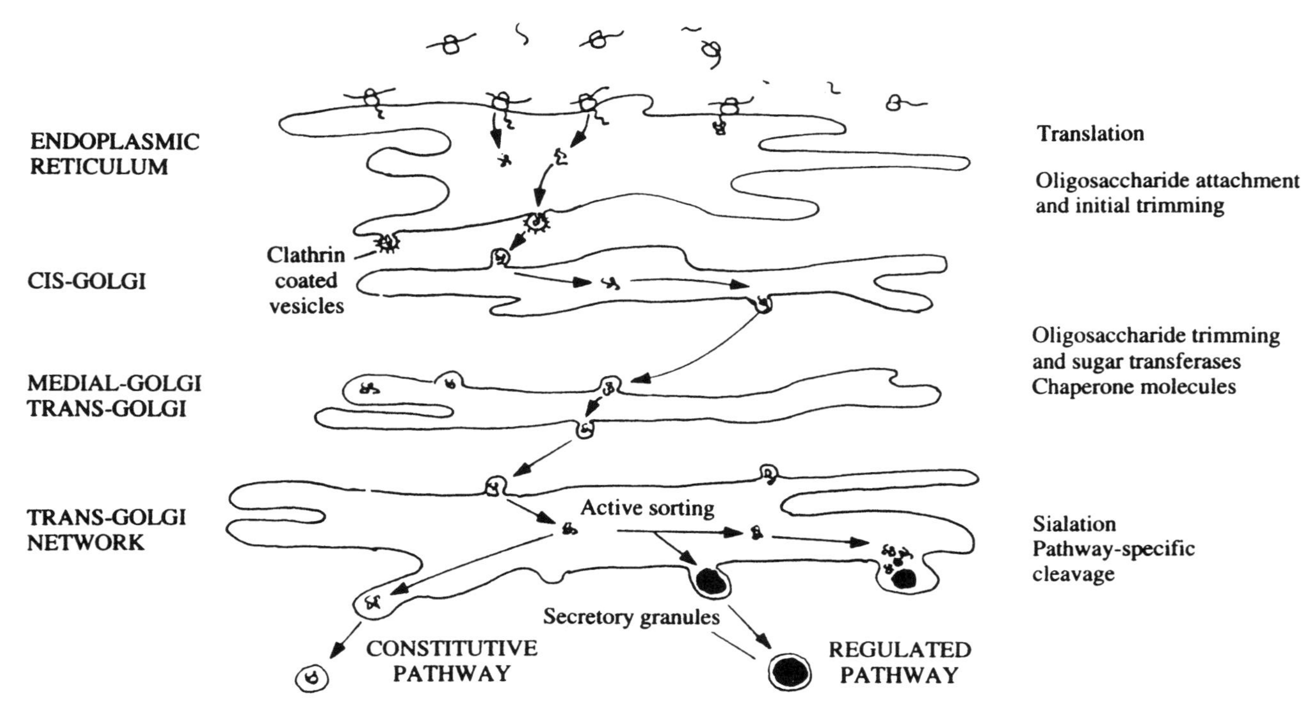

Figure 2.4 Summary view of the post-translational modification and trafficking events that occur as secretory proteins transit the endoplasmic reticulum and the Golgi apparatus before secretion via either the constitutive or the regulated secretory pathway

hanced gene expression, and an understanding of these effects will be critical for full exploitation of production by animal cells.

Many practical examples exist showing that copy number does not always correlate with protein secretion levels in animal cells. Frequently, the most amplified clones are not only no more productive than less amplified clones, but suffer from greater genetic instability and worsened growth and culture characteristics. This is not due to preferential amplification of genomic sites which do not favour transcription because, even when episomally replicating vectors are used, increased copy number does not always give more product. Thus, Jalanko et al (1990) examined the effect of copy number on the expression of tPA in a number of cell types transfected with an Epstein–Barr virus-derived expression vector. Although in some instances copy number paralleled expression levels, in others this was not the case. In particular, K-562 human myelogenous leukaemia cells produced levels of tPA mRNA that were similar to the highest producing cell line studied (a CHO derivative), although tPA production was at least 20-fold less. It was shown that about 96% of the tPA produced in these cells was secreted, so the deficiency in K-562 cells was presumed to be at the level of messenger translation (Jalanko et al, 1990). Clearly, in this system, position effects cannot be responsible for the observed differences in expression.

Observations of this sort suggest that some cells may be poorly adapted to secrete large quantities of correctly processed recombinant proteins. Even CHO cells, which have become the most widely used work horse for protein production, appear to be limited in this way. In an attempt to avoid these limitations, several groups have turned to cell types such as myelomas and other lymphoid cells which may naturally secrete large amounts of protein (typically over 100 μg/ml of immunoglobulin) and which might be considered to be "professional secretors" of protein. It was reasoned that such cells would possess an abundance of all of the post-transcriptional machinery required for efficient translation, processing and secretion.

Another major advantage of such cells is that they grow readily to high concentration in suspension culture and that the requirements for scale up of their growth are well understood. These combined advantages suggest that myeloma cultures may be of particular use for the commercial production of therapeutic proteins, and systems have been developed to exploit this potential. Some examples of this are given in Table 2.4. Several of the systems described use the DHFR/MTX amplification system in lymphoid cells. The study reported by Okamoto et al (1990) is interesting in several respects. First, it shows that effective amplification can be achieved in cells despite the presence of an endogenous DHFR gene. This is important because the development of

Table 2.4 *Production of heterologous proteins using lymphoid cell-based expression systems*

Protein Expressed	Lymphoid Cell Line Used	Promoter/ Enhancer	Amplification	Yield Obtained	Reference
Human monoclonal antibody	P3-X63-Ag-8-6.53 and SP2/O-Ag14 mouse myelomas	Endogenous human Ig promoters	DHFR/MTX*	ca 5 $\mu g/10^6$ cells/day	Nakatani et al, 1989
Tissue plasminogen activator	SP2/O-Ag14 mouse myelomas	Ig promoter/ enhancer	mDHFR/MTX	ca 8 $\mu g/10^6$ cells/day	Hendricks, et al, 1989
Granulocyte-macrophage CSF	Namalwa human ß cell lymphoma	SV40	DHFR/MTX	ca 10-20 $\mu g/10^6$ cells/day	Okamoto et al, 1990
Erythropoietin	Namalwa human ß cell lymphoma	—	—	—	Yanagi et al, 1989
Human monoclonal antibody	SP2/O-Ag14 mouse myelomas	Metallothionein promoter/immunoglobulin enhancer	DHFR/MTX**	20 $\mu g/ml$	Gillies et al, 1989
Lymphocyte surface antigens IL6 IL7	J558L mouse myeloma	Ig promoter	N/A	>100 $\mu g/ml$	Traunecker et al, 1991

* pSV2dhfr was co-transfected with expression vector but no amplification was obtained with MTX

** Amplification of antibody production was obtained without apparent increase in the copy number of the cloned gene.

DHFR⁻ mutants is a long and uncertain process (Urlaub and Chasin, 1980), and no practically useful DHFR deficient mutants of lymphoid cells are available at present. In the study by Hendricks et al (1989), also cited in Table 2.4, use was made of a mutant DHFR gene encoding a DHFR variant with decreased sensitivity to MTX.

A second surprising finding is that the level of amplification of granulocyte–macrophage CSF protein production reported in the Okamoto paper is much greater than the measured amplification of gene-copy number or of specific mRNA (Table 2.5). Thus, in this system, MTX appears to have yield-enhancing effects at the transcriptional and translational levels. The mechanism of these effects remains unknown. Also unexplained is the finding of Gillies et al (1989) on MTX-amplified antibody expression in murine myeloma cells in which antibody protein yield was enhanced by MTX treatment in the absence of detectable gene amplification. Clearly not all aspects of the influence of MTX on the productivity of recombinant proteins are yet understood.

Another way in which high-level expression of protein can be

Table 2.5 *Data from Okamoto et al (1991) showing the gene amplification effects of methotrexate treatment of Namalwa cells transfected with an expression vector containing the genes encoding human GM-CSF and murine DHFR. All the subclones listed were resistant to 800 nM MTX.*

Clone	Fold increase in GM-CSF DNA	Fold increase in GM-CSF mRNA	Fold increase in GM-CSF protein
Parental clone	1	1	1
I-3-1	11	34	700
I-6-6	13	48	590
I-6-2	8	33	400
I-10	21	38	400
I-6-4	18	59	400
I-8	20	68	300

Table 2.6 *Effect of successive cycles of amplification using MTX on tPA yield and doubling time in CHO cells transfected with a plasmid containing the tPA structural gene and a DHFR gene as an amplifiable marker. (Unmodified Bowes melanoma cells included for comparison.)*

Clones	MTX Levels Used (μM)	mg tPA/10^9 cells/24h	Doubling Time (h)
Bowes	0	0.93	24
48501	0.25	6.10	34
48400	2	5.50	30
134003	2	9.8	48
AR9	25	9.4	26
AR14	25	5.25	35

achieved in fermentor-adapted lymphoid cells is by somatic hybridization between high-producing CHO cells and an appropriate lymphoid partner. Hybridomas have been produced from MTX-amplified tPA producing CHO cells and Sp2/0-Ag 14 mouse myeloma cells (Cartwright and Crespo, 1991). In this study, amplification of the CHO cells had resulted in a plateau of tPA production at around 10 μg/10^6 cells/d, and an increase in the doubling time of the MTX-adapted cells (Tables 2.6, 2.7). Hybrids were isolated from a proline-free, MTX-containing selective medium using the auxotrophy of CHO cells for proline and the MTX sensitivity of the myeloma cells as selection pressure. Several hybridomas were obtained which exhibited increased tPA yield and

Table 2.7 *Yields of tPA obtained from several hybriodomas produced by fusion of CHO clone AR9 (Table 2.6) with SP2O/Ag14 myeloma cells*

Hybridoma	Doubling Time (h)	mg tPA/10^9 cells/24h
Parent CHO cell (AR9)	26	9.4
Hyb A8	ca20	10.9
Hyb D6	ca24	16.6
Hyb [illegible]12	ca24	22.0
Hyb B9	ca20	23.2
Hyb F2	ca22	32.6

acceptable growth characteristics (Tables 2.6, 2.7). The best hybridoma obtained, HYB F2, produced 32.6 μg of tPA/10^6 cells/d when adapted to grow in serum-free medium in stirred-tank fermentors. In batch cultures, HYB F2 reached a plateau of tPA production of 100 μg/10^6 cells over 6 d. Under these conditions, tPA was directly visible by HPLC in the unpurified culture supernatant.

Somatic hybridization has also been used by Chenciner et al (1990) to enhance the yield of a recombinant hepatitis-B surface antigen. In this case, the hybridoma was produced from primary monkey hepatocytes and recombinant Vero cells expressing the antigen. Fifty-fold greater expression levels were obtained from the hybridomas than from the unfused Vero cells.

Specific factors and events that can limit expression

Most recently attention has become focused on the specific points in the process of the secretion of recombinant proteins which can limit productivity and how these may differ in various cell types. Results from these studies confirm that cells do indeed have differing characteristic capacities for various processing steps and that cells can be specifically engineered to optimize these capacities.

In a recent study, Pendse, Karkare and Bailey (1992) examined the effect of gene dosage on hepatitis-B surface antigen (HbsAg) expression in amplified CHO cells. HbsAg mRNA level increased with gene-copy number and increased very rapidly as copy number rose above 20. However, the net efficiency of HbsAg expression was reduced in highly amplified clones. Pulse-chase experiments showed that with increasing amplification an increasing proportion of the translated HbsAg fails to be secreted by the cells, suggesting the existence of a bottleneck in the post-translational secretory pathway.

Further investigation showed that intracellular degradation of HbsAg occurred in the highly amplified clones. The authors propose that under amplified conditions a critical post-translational secretory step is rate limiting, and that intracellular proteases are used by the cell to alleviate the block by degrading the accumulated product. In addition to the limitations on protein secretion, this analysis also showed that translational and post-translational events prior to secretion also become overdriven in highly amplified CHO cells.

In other studies on CHO cells it has been reported that there is in general an increase in the synthesis of the primary translation product which parallels the increase in mRNA levels obtained by amplification. However, whereas some proteins such as von Willebrand factor and erythropoietin are efficiently transitted and secreted, others including Factor VIII and non-glycosylated tPA mutants are blocked at the level of transit from the ER (Dorner, Krane and Kaufman, 1988).

In this case, the transit block is apparently due to the stable association of Factor VIII and the tPA mutant form with grp 78, a 78 kDa member of the family of "chaperone proteins" which are localized in the ER. Chaperone proteins are thought to facilitate the folding of proteins destined for secretion. They also appear to be able to retard transit through the ER of proteins that are perceived as being incorrectly folded or aberrant in some other way. Such proteins become sequestered in the ER as a result of binding to the chaperone. Grp 78 has been shown to be identical with the "immunoglobulin heavy-chain binding protein" (BiP) which binds to free immunoglobulin heavy chains in the ER and prevents their secretion until they have associated with immunoglobulin light chains.

Thus, cells engineered to express less of the chaperone proteins might be expected to permit accelerated transit of expressed proteins that would normally be retarded by the chaperone system.

In an interesting use of tandem amplification technology, Dorner et al (1988) introduced a grp 78 antisense expression vector amplified by using the adenosine deaminase/deoxycoformycin system into CHO cells already amplified for tPA production by the DHFR/MTX system. Increasing expression of the grp 78 antisense constructs resulted in a 10-fold decrease in grp 78 expression and a parallel increase in tPA secretion. Another strategy to limit expression of a chaperone protein has been the use of ribozyme technology to block expression of a different protein, grp 94. However, it has become clear from these studies that a certain basal "housekeeping" level of chaperone activity is essential to allow effective monitoring of the processing of endogenous cellular proteins and to maintain cell viability.

The speed and efficiency with which different cell types are able to fold protein correctly is emerging as one of the determinants of protein secretion capacity. In the future, it may be possible to select or to

Table 2.8 *The family of subtilisin-related proteases involved in the proteolytic processing of protein precursors in mammalian cells (reviewed by Smeekens, 1993)*

Protease	Tissue Specificity	Secretory Pathway	Intracellular Localization	pH Optimum
PC2	Neuroendocrine tissue	Regulated	Secretory granules	Acid
PC3 (PC1)	Neuroendocrine tissue	Regulated	Secretory granules	Acid
PACE (furin)	Ubiquitous	Constitutive	Golgi apparatus	Neutral
PACE 4	Ubiquitous	Constitutive	Golgi apparatus	Neutral
PC4	Ubiquitous	Constitutive	Golgi apparatus	Neutral

engineer cells with the ideal chaperone molecule complement for maximizing expression of a given recombinant protein.

The secretion of proteins that have been bound to members of the grp group of chaperones is an energy-requiring process, and the more stably bound proteins require the consumption of more ATP for transport. It has been proposed that the intracellular availability of ATP may be a limiting factor for the release and secretion of such proteins (Dorner, Wasley and Kaufman, 1990).

Processing enzymes

Many proteins require specific and often sequential proteolytic clipping as part of the intracellular processing sequence before efficient secretion can be achieved. These include virtually all peptide hormones and neuropeptides, many growth factors and growth factor receptors and a large proportion of plasma proteins. A newly discovered family of calcium-dependent serine proteases which resemble bacterial subtilisin in their domain structure has now been shown to be responsible for this cleavage (Barr, 1991). All these enzymes exhibit similar specificity and cleave at dibasic sites within the protein precursors (reviewed by Smeekens, 1993). However, the various enzymes show tissue specificity and are associated with different secretory pathways (Table 2.8).

In cells engineered for high expression of recombinant proteins, the endogenous processing proteases may not be able to keep up with the expression of precursors which may then accumulate intracellularly because secretion cannot take place (Bloomquist, Eipper and Manns, 1992). Cells engineered so that they also exhibit enhanced processing protease expression should be able to achieve proteolytic processing at an increased rate and so increase secretion capacity (Rehemtulla, Dorner and Kaufman, 1992). This approach is currently under investigation.

It has also been suggested that enzymic glycosylation steps occurring in the Golgi apparatus may become rate limiting when translation is being driven hard in heterologous host cells. Attempts have been made to improve product yield by increasing the oligosaccharide synthesis capacity of the cells by expressing increased levels of glycosyl transferases, but significant gains in production have not so far been reported (Lee, Roth and Paulson, 1988; Nakazawa et al, 1991).

Uncoupling cell growth and secretion

Most of the cell lines currently used in production secrete proteins continuously via the constitutive secretory pathway as they are synthezised. However, most differentiated cells from endocrine and exocrine tissues, whose natural function is protein secretion, make use of the constitutive secretory pathway (Figure 2.4) in which proteins are sequestered in storage granules and are released only when the cells are stimulated with a secretagogue (Kelly, 1985). Grumpp and Stephanopoulos (1991) have used this phenomenon to optimize the yield of proteins from At-20 rat pituitary tumour cells expressing human pro-insulin and human growth hormone. The cells were cultured in serum-containing growth medium, during which time product accumulated in storage granules, then they were put onto serum-free medium and treated with a secretagogue to stimulate release of the accumulated product over a 2-hour period. It was claimed that this 'controlled secretion' approach favours complete post-translational processing of the proteins (which occurs inside the storage granules), and also permits concentrated product release in low-protein conditions without the need to expose the cells continuously to a potentially damaging low-protein environment. Multiple cycles of product accumulation and release are said to be possible (Grumpp and Stephanopoulos, 1991), although it is not yet established that the process could be operated industrially.

Improving translation efficiency

The process of translation may itself become limiting when large amounts of messenger are produced. In some cases, availability of trans-acting intracellular proteins such as initiation factors can represent a bottleneck. Initiation factor activity is modulated by their phosphorylation state. Thus phosphorylation of eIF-2 alpha inhibits protein synthesis by preventing the recycling of this factor whereas phosphorylation of eIF-4F has the opposite effect and increases protein synthesis. Several research groups are investigating the possibility of enhancing protein synthesis by manipulating these effects. One approach has been to introduce a mutant eIF-2 alpha which cannot be phosphorylated in the hope that this will result in continuing enhanced translation. It remains to be seen whether such manipulations of factors that are so central to the control of cellular metabolism will result in a practical system for obtaining improved recombinant protein production.

Thus, at our present state of knowledge, it may not be the most cost-effective solution to concentrate all development efforts for the production of proteins from recombinant animal cells on systems that contain the maximum possible copy number of the required gene because instability of the clone, poor cell growth and inefficient processing may all result, and these deleterious effects may outweigh some of the advantages of high copy number.

Clearly, to be able to drive production from cells optimally it is also essential to have a better understanding of the regulation of cellular metabolism and of the mechanisms of synthesis and secretion of product. It is towards an enhanced understanding of these parameters that many research groups are currently striving.

3 Generation of Biomass

As discussed in Chapter 2, specific protein production (pg/cell/hour) can be dramatically increased by the creation of more efficient producer cells which are capable of high-level transcription coupled with fast and accurate processing and a secretory capacity of matching performance.

However, producing commercial quantities of recombinant protein even by the most productive cells still requires large-scale cell cultures. This need has led to the rapid and continuing development of new approaches to mass cell culture and has been instrumental in dispelling certain dogmas concerning the impracticality of commercial production from animal cells. It has also led to an ongoing re-evaluation of product generation from animal cells according to more sound bio-engineering principles.

Small-scale, low-density animal cell cultures in classical serum-containing medium are relatively simple systems to operate. Since the metabolism of the cells is slow, oxygen demand can be satisfied by simple diffusion or slow stirring, pH changes are slow and easily controlled by buffering and the effect of the production of toxic substances and of proteases are largely abrogated by the absorption and inhibitory capacities of serum.

This situation changes dramatically when large quantities of cells are cultured intensively for the industrial production of kilogram quantities of recombinant proteins and all of these parameters are pushed to new limits. This chapter considers the progress that has been achieved in the key areas to satisfy these demands and discusses the problems that remain and some possible solutions.

Media for animal cell culture

Major changes have occurred and are continuing to occur in the nature of the medium used for cell culture. One of the primary perceived difficulties in the culture of animal cells, referred to in Chapter 1, has always been the complexity of the medium used for growth and for product generation. Complexity implies high cost and low reproducibility, factors which are incompatible with the demands of industrial processes. Much has been accomplished to improve these aspects of medium by optimizing the components.

Table 3.1 *Disadvantages of use of serum in medium*

*	High cost of quality sera - especially FBS
*	Inherent variability of different serum batches
*	Possibility of contamination: viruses, mycoplasma, bacteria, fungi, endotoxin
*	High protein content may complicate down-stream processing procedures.

Serum

Early studies with animal cells in culture used media which were essentially isolated body fluids such as lymph or plasma. In the 1950s, the emerging vaccine industry began to demand large-scale animal cell cultures for the first time. The need to standardize vaccine production led to the first attempts to produce culture medium of defined composition. Eagles's medium, produced in 1959, was one of the first attempts to approach this problem systematically. However, even when the cells were fed with the complex mixture of amino acids, minerals, vitamins and carbohydrate described by Eagle (1959), it remained necessary to supplement the medium with up to 20% of animal serum to achieve satisfactory growth. Because of its rich content of growth-stimulating factors and its low gamma globulin content, foetal bovine serum became the standard supplement for tissue culture medium and is routinely employed at a concentration of 10%. The majority of established animal cell biotechnology-based production processes currently use media supplemented with animal serum. However, the use of such serum poses a number of difficulties which impact the safety, reproducibility and cost of the production of pharmaceutical proteins (Table 3.1). These difficulties can be minimized by using serum from properly controlled sources. They include high cost and uncertain supply (particularly in the case of foetal bovine serum) and the fact that batches may vary appreciably in their growth-promoting capacity. Batch variation may be due to the inherent differences between animals and to the effects of variations in their environment, particularly such factors as the time of year at collection, feed stuffs used and the effect of local climatic conditions (especially drought). In addition, different cell types may respond quite differently to different serum batches for reasons that are rarely understood.

These problems have led to the standard practice of 'batch testing' in which several lots of serum are placed on reserve with a producer while they are tested extensively (and expensively) by the user before selection of the highest-performing material for the specific operation in question.

Serious serum producers have invested very considerably to develop a reliable and validated production process by standardizing conditions of animal husbandry, by careful monitoring of the health of animals used for collection and by improving processing technology. Considerable success in maintaining quality and in limiting batch-to-batch variation has been achieved. However, despite all the efforts to standardize conditions of collection, foetal bovine serum remains a by-product of the dairy industry whose availability is determined by current dairy policies which in turn determine the number of pregnant cows going to slaughter. For example, foetal bovine serum availability increased suddenly in 1986 when large numbers of dairy cows were slaughtered in the United States due to a revision of the structure of dairy subsidies. Prices of foetal bovine serum fell sharply accordingly, but prices of foetal bovine serum of approved origin are now rising steadily as current agricultural practices are again limiting the number of pregnant cows going for slaughter.

The possibility of contamination by adventitious agents is an ever-present threat when using animal sera or other animal products, and extensive testing is required by the regulatory authorities to show the absence of contaminating viruses or other microorganisms and of endotoxin. Much effort by both suppliers and users is devoted to assuring that such contamination does not and cannot occur. On the part of the supplier this requires rigorous health checks on the animals employed, Good Manufacturing Practice (GMP) facilities for the collection and processing of serum, rigorous documentation and batch traceability and thorough quality-control testing to assure that the serum offered satisfies regulatory requirements. Quality-control (QC) testing usually takes the form of extensive virological and microbiological testing of the product (in addition to growth-performance testing) after application of a validated processing and sterile filtration procedure. Filtration increasingly involves passage through several 0.1 μm filters.

For these reasons correctly processed serum for cell culture is an expensive commodity. It is also a commodity whose availability and, hence, price, is affected by shifts in national and international agricultural policies. These facts are not understood by many serum users. From the serum user's point of view, quality assurance of serum will involve auditing the supply operation and probably also more or less extensive additional tests within their own quality-control facilities.

The quality control applied to collection and processing of serum is critically important at every step. In addition to its obvious bearing on microbial sterility, endotoxin content and contamination by exogenous viral agents, the quality of the processing has other direct effects on the virological status of the serum. As an illustration of this, the case of bovine viral diarrhoea (BVD) can be cited. BVD is one of the panel of bovine viruses which must be definitively excluded from serum used for cell culture-based pharmaceutical production. It is generally accepted that probably all bovine populations contain BVD which is transmitted transplacentally and most commonly causes clinically inapparent infection (Roeder and Harkness, 1986). In serum production from such a population, the level of BVD in the unfiltered serum product depends heavily on the quality of the processing steps used. BVD occurs intracellularly in leucocytes and is released by cell lysis. Thus, effective centrifugation procedures to remove all leucocytes are very important, as is rapid processing to limit leucocyte lysis. Careful attention to these points are essential for producing an end-product which is truly BVD-free.

The wisdom and necessity of comprehensive testing for viruses, and the shortcomings of testing, were graphically illustrated recently by the bovine spongiform encephalopathy (BSE) outbreak that was centred in Great Britain. In this instance, a newly described transmissible agent which causes brain degeneration was discovered in the British bovine population. Subsequent studies showed that the agent belongs to the group known as prions or viroids (Collee, 1991) which are both extremely small and extremely resistant and which cannot therefore be reliably eliminated by filtration or by the standard virus inactivation processes such as heat inactivation, acidification or gamma irradiation. The BSE agent is transmissible to other species and presents obvious risks if it is present in culture medium used for the production of pharmaceuticals. In consequence, the discovery of BSE entirely destroyed the market for serum from all countries, particularly the United Kingdom, where cases of the disease are known to have occurred. This added further to the difficulty and cost of obtaining pharmaceutical grade foetal bovine serum.

The BSE situation is the latest and most extreme example of the problem that even the most rigorous test procedures cannot absolutely guarantee that any given agent is completely absent from a batch of serum. As discussed later, properly validated virus inactivation procedures could achieve this objective, but most procedures which inactivate viruses also damage the cell-growth promoting capacity of serum, as well as adding substantially to its cost. For this reason, the policy of specifying that only serum from specified countries of origin where the virus in question is thought not to occur has become widely required by regulating agencies. This has resulted in the creation of a league table of acceptable geographical sources with New Zealand, the United States and Australia occupying the highest places.

Not surprisingly, these restrictions further limit availability and in-

Table 3.2 *Virus clearance from BSA by the HAMOSAFE process*

Spiked Virus	Pretreatment Concentration	Model (Cell Line)	Measured Clearance Factor*
Scrapie/BSE**	2 x 10E8	Mouse	>10E5
BVD	1 x 10E8	BHK	3 x 10E6
IBR	1 x 10E5	BHK	7 x 10E4
PI3	80 HTH units	?	—
FMD	2 x 10E7	MDCK	1 x 10E6
MVV	1.2 x 10E9	WSCP	3 x 10E6
Orf	2 x 10E9	PAL-6	3.4 x 10E6

Source in data presented at German ETCS Meeting, Biberach an der Riss, 26 May 1992

* In all cases, no virus was detected after treatment. The clearance factor given reflects the experimentally determined detection limits in the relevant test system.

** Virus Codes:

BVD - Bovine Viral Diarrhoea
IBR - Infectious Bovine Rhinotracheitis
PI3 - Parainfluenza 3
FMD - Foot and Mouth Disease
MVV - Maedi-Vissna

crease the cost of correctly collected, processed and validated serum. An indication of the impact of this series of restrictions on the availability of serum for pharmaceutical use is given by the development of a major black market industry in which serum batches of dubious origin and quality have been falsely represented regarding both their country of origin and the quality control applied to them (Hodgson, 1991; Hodgson, 1993). The picture may now be changing, however, following a recent exciting development in BSE control. The Austrian company Hamosan has developed a fully validated BSE inactivation process that can be applied to many proteins without causing loss of biological function. This Hamosafe process has been most extensively studied with bovine serum albumin (BSA), on which it was shown to reduce BSE titres to undetectable levels in samples of BSA spiked with high titre BSE (Table 3.2). More recently the Hamosafe procedure has been shown to be effective when applied to foetal bovine serum, to bovine transferrin and to several purified enzymes (H. Reichl, personal communication, 1992). The best available information indicates that BSE does not occur at detectable levels in correctly collected serum (infectivity titre <10 even in serum collected from animals showing high BSE titres of up to 10^9 in neural tissue; Darling et al, 1992). The Hamosafe process therefore offers considerable reassurance when applied to serum and purified serum proteins intended for use in tissue culture.

Donor serum

One approach to limiting the risks associated with animal serum has been the donor-serum concept in which herds of animals are assembled exclusively for serum production and are isolated from other animals. The herd is constantly monitored to ensure the absence of specified viruses, and all new animals are rigorously screened before they are allowed to enter the herd. Each animal is individually identified, and comprehensive batch records assure that serum can be traced to a particular animal. Animals are bled at intervals by aseptic venepuncture under perfectly controlled conditions. Advantages of this system over blood collection at slaughter are obvious and include

- better control over animal husbandry.
- complete knowledge of the health (especially virological) status of all animals used.
- complete control of the blood collection conditions, permitting truly aseptic collection and the use of a rapid, fully integrated GMP process from collection to sterile filling.
- reproducibility of batches since the herd is maintained over long periods.
- full traceability to the source animal with the possibility that the user can specify which animals should be used to assemble a batch.

Donor serum is currently available from horses and from calves. In many applications, donor serum is as effective in cell-growth promotion and in product generation as foetal calf serum. For some applications, notably monoclonal antibody production from hybridomas, horse serum may be superior in some cases. TCS Biologicals Ltd. currently maintain probably the largest herd of horses for GMP quality donor serum production in Europe, in excess of 600 horses. TCS Biologicals Ltd. produces donor horse serum that is specially tested for high performance in tissue culture and sold under the name of Equicell.

In addition to offering better protection against infectious agents and better standardization of the product, donor serum is cheaper than foetal bovine serum. In consequence it should be evaluated for those production processes whose medium supplementation with serum is essential. Overall, production of serum for pharmaceutical use is complex both technically and administratively. Although batches of serum that are satisfactory for 'one off' use, can be found relatively easily on the spot market, industrial protein production using serum can be effectively achieved only if the serum user works with a serum supplier whose discipline and consistency of manufacture permits regular delivery of validated serum which has reproducible performance and is of impeccable pedigree and traceability.

Serum-free media

Many, perhaps most, of the established mammalian cell-based production plants operating today employ serum medium supplements despite the weight of disadvantages that this imposes (Table 3.1). One major reason why this situation persists is that the process specifications for many of these operations were finalized when production-tested serum-free media were a less realistic proposition than they are at present. Conversion of such an operation from serum use to a serum-free system would pose many great technical and regulatory difficulties. However, it is now quite apparent that an appropriately designed serum-free medium offers major advantages at all stages of bioproduct generation and that any new process now being designed should be conceived from the outset for serum-free operation.

Several advantages of using serum-free media are immediately apparent from the preceding discussion of the problems due to biological variability encountered when serum is used, but the advantages of serum-free medium extend beyond the cell growth and productivity aspects of the biotechnology process. Perhaps the most important advantage in the overall cost of the process is the simplification of downstream processing that can usually be achieved. Since downstream processing can account for 70–80% of the total process costs, any gains in time or yield in this area are of major economic importance. Indeed, the use of serum-free medium can be essential in some cases in which low levels of the required product are produced because it may be impossible in such cases to devise an economically practical purification scheme to recover the product from the high concentration of complex proteins which serum brings to the medium (Table 3.3).

Historically this problem has been partially combatted by dividing the production operation into the cell growth phase and the product-generation phase. Where necessary a serum-rich medium can be used for cell growth which is replaced by a production medium containing reduced serum for the product-release phase. However, low levels of secreted protein are difficult to purify, even from medium containing only 1% serum, and the required medium change adds to both cost and risk by complicating the process. Clearly, this problem is avoided if a single serum-free, or better still, a protein-free medium can be developed which can be used throughout the production process.

Another important use of serum-free medium is in research aimed at the characterization of the growth-factor requirements of a particular cell type. This cannot be done in the presence of serum because serum

Table 3.3 *Generally accepted concentration ranges for the major protein components in animal sera for cell culture*

Protein (g/100ml)	Foetal Bovine Serum	Donor Calf Serum	Donor Horse Serum	Human Serum for Comparison
albumin	2—3.5	2.5—5	3—4.5	4—5
alpha globulins	0.7—2	0.7—1.5	0.7—1.4	0.2—0.6
beta globulin	0.4—0.9	0.5—1.5	1—2.3	0.2—0.5
gamma globulin	<0.02	0.5—2.2	0.3—1.5	0.3—0.8
haemoglobin	<0.03	<0.03	<0.03	—
Total Protein	3.5—5.5	5—8	5—7.5	6—8

Figures are a consensus from several manufacturers' specifications. All figures are in g/100 ml.

Table 3.4 *Advantages and disadvantages of using serum-free medium*

Advantages	Possible Disadvantages
* No risk of viral contamination (if no animal-derived additives are used)	* Fastidious nature of cells - may require a custom made medium for each cell type
* Constancy of medium composition	* Possible slow adaptation to growth in serum-free conditions
* Security of supply	* Loss of positive effects of serum:
* Definition of the nutritional environment	protease inhibition
* Greatly facilitated downstream processing	toxin neutralization
* Cost savings at high volume	buffering capacity
	mechanical protection

contains a complex and undefined cocktail of growth factors. Although such studies are not directly involved in bioproduct generation, they represent a critical step in the design and optimization of serum-free medium for specific producer cells. The major benefits and some possible disadvantages of serum-free medium use are summarized in Table 3.4.

Medium design

A logical extension of the processes by which vectors are engineered for optimum product yield and efficient fermentor configurations are developed for maximum cell yield is the design of a medium which is ideally suited to product generation from the chosen producer cell in the required fermentor conditions. However, design of a defined serum-free medium is not a simple undertaking. Serum is an effective growth supplement for most cell types because of the complexity and the mul-

Table 3.5 *Multiple nutritional and protection factors supplied by serum*

Factor	Effective concentration
Specific growth factors (see table 3.6)	1—100 ng/ml
Trace metals:	
Iron	1—10 μM
Zinc	0.1—1 μM
Selenium	0.01 μM
Lipids:	
Cholesterol	ca10 μM
Linoleic Acid	0.01—0.1 μM
Polyamines:	
Putrescine	0.01—1 μM
Ornithine	0.01—1 μM
Spermidine	0.01—1 μM
Attachment Factors	—
Mechanical protection	—
Buffering Capacity	—
Neutralisation of toxic factors	—

tiplicity of growth-promoting factors, nutritional factors and cell-protective agents that it contains assure that it can satisfy the needs of practically any type of cell. (Table 3.5)

This complexity is necessary because, unlike microbial cells, animal cells require more than just simple nutrients for growth and proliferation. In particular many cell types have an absolute requirement for members of the group of small (5–30 kDa) proteins known as growth factors (Table 3.6). Such factors are needed for cell growth but may also be necessary for the maintenance of a productive phenotype and for product generation. In other cases (Williams et al, 1990; Strasser et al, 1991), the continued presence of growth factor is essential throughout the life of the cell and removal or absence of the factor initiates the pre-programmed autodestruction of the cells by a process called apoptosis which is distinct from the general degenerative processes initiated by nutrient deprivation or by trauma.

Growth factors function by occupying specific receptors on the surface of the cell, resulting in the generation of signals within the cell which govern proliferation and the expression of a particular phenotype. Different cells have different growth-factor requirements (see Table 3.6), and serum batches may contain different relative concentrations of the factors. In addition, growth factors which are stimulatory for one type of cell may inhibit the growth of others (Moses et al, 1987).

Most serum-free medium formulations contain low levels of selected growth factors. Insulin and/or insulin-like growth factors are probably the most widely used because of their general stimulating effect on the uptake of glucose and amino acids, and on the metabolism of these nutrients. Other factors are sometimes added which are more or less specific for the required cell type (Table 3.6).

For some cell types that normally grow attached to solid substrates, serum is also a critical source of protein attachment factors such as fibronectin. However, many cell types secrete their own extracellular matrix proteins once they are established in culture. Many of the growth factors contained in serum bind more or less avidly to matrix proteins, and this dynamic interplay between cell receptors, extracellular matrix and growth factors is also thought to be a major factor in the control of cell growth and maintenance (Gospodarowicz et al, 1987).

Purified fibronectin and other matrix proteins from animal sources have been used to supplement the medium for cells which have an absolute requirement for attachment factors. However, this again raises the possibility of contamination by animal-derived adventitious agents and imposes the need for careful biological screening. A recent development is the production of a genetically engineered synthetic cell-adhesion molecule called Pronectin® which provides multiple copies of the RGD adherence motif in a designed polypeptide sequence. This molecule, available from Protein Polymer Technologies Inc., San Diego, California, is reported to be more stable than fibronectin and to have improved cell-adhesion properties.

In addition to the protein growth factors, serum is an important source of lipids and trace elements for cellular metabolism, and serum proteins are involved in the transport of these molecules and their presentation to cells. Thus, serum albumin, in particular, binds and transports many substances, including steroids and other lipids, mainly fatty acids. This protein-bound form of presentation is important since free lipids are often poorly soluble in aqueous media, and free fatty acids may be toxic to cells in culture.

Purified bovine albumin is sometimes added to medium as a lipid source. It should be noted, however, that several methods of purification are used commercially to prepare albumin and that the method used greatly affects the amount of lipid that remains bound to the protein. In general, ethanol and other solvent fractionation methods tend to strip lipids from the albumin more than the chromatographic procedures also in use. There are also reports that chromatographically purified albumin retains more of the native albumin conformation than solvent-purified material and that this also favours cell growth.

Although albumin can be a useful supplement in serum-free medium, it should be noted that when albumin from bovine blood is used the

Table 3.6 *Growth factors*

Growth Factor	Molecular Weight (kDa)	Original Source (recombinant material now usually used)	Effective Concentration Range	Cell Types Stimulated	Receptor
Insulin	5.7	Pancreas	1-10 μg/ml	Anabolic and mitogenic for many cell types	Tyrosine kinase linked
Insulin-like growth factors (IGF) (also called somatomedins)				Similar to insulin but much more potent. Wide range of cell types including fibroblasts, chondrocytes, myoblasts, neuroectoderm derived cells	IGF 1 tyrosine kinase linked (also binds IGF 2 and, to a lesser extent, insulin). IGF 2 has a specific receptor which is not tyrosine kinase linked
IGF 1	7.6	Rat liver	5-40 ng/ml		
IGF 2	7.5	Rat liver	5-25 ng/ml		
Nerve growth factor (NGF) (several classes exist)	26.5	Male mouse sub-maxillary gland	5-20 ng/ml	Neuroectoderm derived cells	Non-tyrosine kinase linked receptor
Epidermal growth factor (EGF)	6	Male mouse sub-maxillary gland	1-20 ng/ml	Potent mitogen for a wide range of ectoderm and endoderm derived cells	Two classes of receptor with intrinsic tyrosine kinase activity
Transforming growth factor (TGF)				Mitogenic for EGF sensitive cells. Has 30% homology with EGF and binds to EGF receptor with equal potency	Binds to EGF receptor
TGF alpha	7.0		5-50 ng/ml		
TGF beta 1	25.0	Platelets	0.1-3 ng/ml	Structually distinct from TGF alpha. Affects virtually all normal and neoplastic cells - activity depends on cell type. Stimulates the growth of mesenchymal cells in soft agar in combination with EGF and TGF alpha. Inhibits growth of many cell types and many cellular activities such as enzyme and growth factor release, differentiation, matrix formation and resorption, etc, etc.	Ubiquitous receptor
TGF beta 2	25.0		0.1-3 ng/ml		

Fibroblast growth factor (FGF) Acidic FGF (aFGF) Type I & Type II Basic FGF (bFGF) Type I & Type II	 16.4 & 17 6.2 & 7.2	 Bovine Brain Bovine Brain	 10-100 ng/ml 0.1-10 ng/ml	Mitogenic for most mesoderm and neurectoderm derived cells and some epithelial cells including fibroblasts, endothelial cells, hepatocytes oligodendrocytes, and chondrocytes. Also BHK21 cells, CHO cells, Balb/C cells & 3T3 cells.	aFGF and bFGF probably share the same protein kinase C-linked receptor
Platelet-derived growth factor (PDGF)	28-36	Platelets, endothelial cells	1-20 ng/ml	Potent antigen for a wide range of cells of mesodermal origin including fibroblasts, smooth muscle cells, glial cells	PDGF receptors with tyrosine kinase activity.
Interleukin 6 (IL6) [interferon β2]	20.6		1-10 ng/ml	Mitogenic for hybridomas and stem cells. Stimulates maturation and IgG production by B cells; IL2 production and T cell growth.	

same concerns as for serum regarding adventitious agents, especially BSE, must be addressed. Also, reasonably priced albumin preparations may not be sufficiently purified with respect to other serum proteins and may therefore complicate product recovery. An important example of this occurs in systems in which protein A- or protein G-affinity chromatography is used for immunoglobulin purification from culture supernatants, in such cases residual gamma globulin in albumin preparations can pose problems by being copurified with the required antibody and by reducing the usable capacity of the chromatographic support.

Systems have been devised to avoid the use of albumin as a lipid carrier in culture medium by presenting lipid to cells in the form of stable micro-emulsions of the type used clinically for parenteral nutrition and also by using wholly synthetic hydrophilic carrier molecules such as cyclodextrins (Yamane et al, 1982).

The lipid composition of cell culture medium is a factor of considerable importance because the lipid composition of animal cells varies considerably, depending on the nature of the lipids supplied in the culture medium. In particular, the fatty acyl components of membrane phospholipids can rapidly become modified with subsequent effects on cell-membrane fluidity and transport properties. The impact of this on cell growth and on product generation is still poorly understood at present, but serum-free medium should contain a source of lipids and this component of medium is one of the important design parameters yet to be optimized (Rosenthal, 1987).

Beyond the metabolic and nutritional effects, serum proteins also possess physicochemical properties which favour cell growth and product generation. These include protection against fluctuation of pH by acting as an effective pH buffer, absorption and neutralization of toxic compounds such as heavy metals, proteolytic enzymes and endotoxin and mechanical protection against the shear forces that can be encountered in stirred and sparged cultures (Table 3.5).

The mechanism of the shear protective effect of serum and of proteins such as serum albumin remains unclear. However, it is established that the effect is at least mainly mechanical rather than metabolic since the protective effect occurs immediately after addition of the protein to serum-free medium (Van der Pol and Tramper, 1992). The fragility of animal cells in fermentors will be discussed in the section on fermentor design.

Obviously one of the major difficulties in the design of a serum-free medium is to find a simplified formula able to replace serum in all of these diverse roles in a way that is satisfactory for the type of cell in question. Attempts to achieve this have resulted in some very complex medium formulations which in some cases still contain high levels of added protein.

In fact, most cell types have quite specific requirements for a limited number of the growth factors found in serum, although this specificity may not be apparent when cells are cultured in the rich soup of factors that serum provides. A given cell type therefore requires the development of a specifically optimized medium before satisfactory growth and productivity can be achieved. This has led to the commercial introduction of new serum-free medium products for the culture of widely used specialized cell types such as hybridomas or CHO cells (Table 3.7). In cases such as these, in which the nutritional requirements of the cells are becoming well understood, medium formulations are now becoming much more simplified. In several cases, truly defined media which contain no proteins, peptides or animal extracts are now available for use with particular cell types (Table 3.7). The actual cell line used in production by a pharmaceutical manufacturer would very probably benefit from fine tuning specifically for that cell line. This can be done in-house, but it is a time consuming exercise, and it may not be cost-effective for a company to maintain the necessary expertise for an operation which may be necessary only infrequently when a new production cell line is established. On this basis, several of the serum-free medium manufacturers offer contract optimization of medium to suit the producer's own cell line (Jayme and Greenwold, 1991).

Engineering a new medium

The design of a new serum-free medium is still basically empirical. The most widely used approach is to progressively lower the serum concentration in cell culture and to determine which factors are needed to restore growth and product generation capacity. This lengthy process permits the identification of critical protein growth factors. The growing data base for the basic cell lineages, epithelial, mesenchymal and so on, provides some guidance for appropriate starting points. Several special nutritional requirements have been identified and can be applied generically. For example NS-1 myelomas are strict auxotrophs for cholesterol (Myoken et al, 1989) and it is to be anticipated that hybridomas derived from NS-1 might also require the addition of cholesterol to the medium. In general, transformed cell lines generally have lower growth-factor requirements than untransformed cells. In some cases protein can be completely eliminated from media for transformed cells (see Table 3.7).

For simple low molecular weight nutrients, the availability of rapid analytical methods, notably HPLC, has made it possible to analyze the

Table 3.7 *A selection of serum-free media currently available*

Serum Free Medium Supplier	Cell Types				
	Hybridoma	CHO	Insect Cells	Lymphoid Cells	General
Bio-Whittaker Inc.	Ultradoma (30) Ultradoma PF (0)	Ultra-CHO (<300)	Insect Xpress (0)	Ex-Vivo range (1000-2000)	UltraCulture (3000)
Boehringer Mannheim	Nutridoma range (40-1000)	—	—	—	—
Gibco	Hybridoma SFM (730) Hybridoma PHFM (0)	CHO-SFM (400)	SF 900	AIM V	—
Hyclone Laboratories	CCM-1 (210)	CCM5 (<400)	CCM3 (0)	—	—
ICN Flow	Biorich 2	—	Biorich 2	—	—
JRH Biosciences (Seralab)	Ex-cell 300 (11) Ex-cell 309 (10)	Ex-cell 301 (100)	Ex-cell 401 (0)	Aprotain-1 (0)	—
Sigma	QBSF 52 (45) QBSF 55 (65)	—	SF insect Medium (0)	—	QBSF 56 (430)
TCS Biologicals Ltd.	SoftCell-doma LP (30) SoftCell-doma HP (0)	SoftCell - CHO (300)	SoftCell-insecta (0)	—	SoftCell - Universal (3000)
Ventrex	HL-1 (<50)	—	—	—	—

Protein content in μg/ml is indicated in brackets where known consumption during culture of specific metabolites such as amino acids. It is then possible to optimize medium composition for a given cell type by supplementing rapidly depleted components and perhaps also by reducing the concentration of under-utilized components. However, interpretation of such studies is complicated by the dynamic interaction between the utilization of various medium components. This has led several groups to base optimization studies on factorial experimental design aimed at accommodating complex systems involving multiple interacting factors (Plackett and Burman, 1946; Mignot et al, 1989). Additional complications can arise because cellular metabolism can change during the course of a fermentation, and different amino acids may become critical at different phases of the culture (Bell et al, 1991). Excessively reductionist approaches to medium design should therefore be applied with due caution.

Difficulties encountered with serum-free media

Cells grown in serum-free media are not cosetted by all the beneficial effects of serum listed in Table 3.4. This fact manifests as an increased fastidiousness of cells and a reduced tolerance to variations in medium quality and perturbations in culture conditions. As we have seen, a dedicated medium usually has to be developed specifically for each cell type, and this fastidiousness is particularly apparent when it is necessary to cultivate a differentiated cell type.

Cell growth conditions become more critical in the absence of serum. For example, minor fluctuations of pH which could be tolerated in serum-containing medium are less tolerated in serum-free conditions. Tighter control of the key process parameters is therefore required.

The loss of other serum protective effects may also be significant. In general, better quality reagents, particularly water, are required since the capacity of serum to neutralize toxic substances is no longer available. In the absence of serum, even closer attention to the elimination of endotoxin from equipment and reagents is necessary for the same reason. Protection against shear forces in fermentors is also reduced in the absence of serum proteins. Several medium additives have been used to overcome this problem, the most widely used being nonionic polyol surfactants such as Pluronic F68 (Mizrahi, 1975; Handa-Corrigan et al, 1992).

Another possible inconvenience when switching from serum-containing to serum-free medium is the long adaptation time or weaning period required by many cells. Several weeks at least may be required for cells to adapt fully to a new set of medium conditions. This underlines the investment in time that a programme to design a new serum-free medium from scratch is likely to require (see also Table 3.4).

Finally, it has still not proved possible to develop effective defined or serum-free medium for some critical operations. An important example of this fact is in the early stages of hybridoma culture in which hybridoma production and cloning-out operations still require the use of feeder layers or of specially conditioned medium to provide essential but uncharacterized growth factors. (Suitable conditioned media for use as hybridoma growth supplements include Condimed H1 from Boehringer Mannheim, Briclone from BioResearch Ireland and Hybridoma Enhancing Supplement from Sigma.) Once past this initial critical stage, hybridomas can usually be readily adapted to and grown in serum-free media.

Despite the difficulties of serum-free medium development and use, these products represent a major growth area in the biotechnology industry as indicated in Table 3.7. An holistic evaluation of most biotechnology production processes in animal cells clearly indicates that serum-free medium and, ultimately, protein-free medium, is the most cost-effective production route. As a general strategy, all new production should be designed on this basis.

The medium production industry is accordingly geared for industrial scale production and offers large batch sizes (tens of thousands of litres) and large containers adapted to the fermentors and other operating plant of the user. Recent developments have been the production of complete powdered media which only require rehydration on site with tissue culture quality water, and 50x concentrated complete media from which ready-to-use medium can be simply generated by on-line dilution equipment (Jayme and Greenwald, 1991).

Fermentor design for animal cell cultures

If increased biomass is needed to generate sufficient product, it can only be generated in one of two ways, each of which has its protagonists and its opponents. Either the size of the fermentor must be increased in proportion to the increase in demand or the configuration of the fermentor must be altered so that a much higher cell density can be achieved and maintained.

Until recently, animal cells have been grown as low-density monolayer cultures or in stirred-tank fermentors that were adapted versions of the well-understood fermentation equipment used for microbial cultures. However, the special requirements for effective growth of animal cells and the increased demand for their products on an industrial scale have caused much ingenuity to be directed to the design of new fermentor configurations specifically devised for the culture of animal cells at greatly increased cell density. Figure 3.1 gives an overview of the

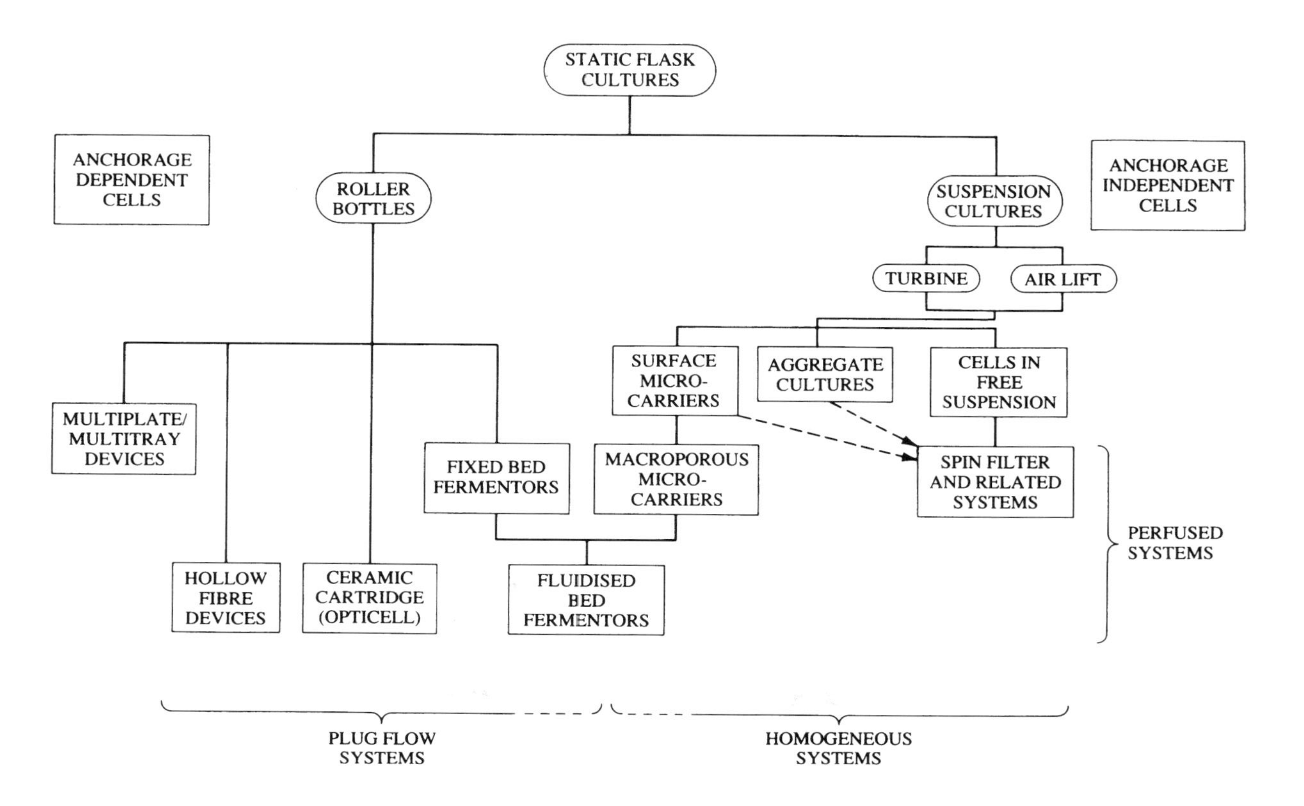

Figure 3.1 Overview of the evolution of the different systems available for scale-up of animal cell cultures

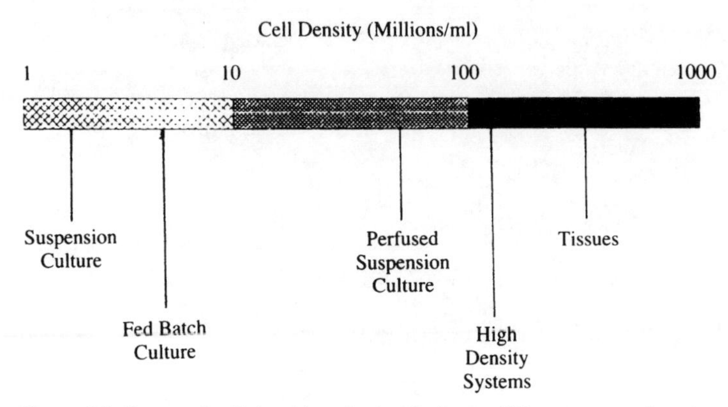

Figure 3.2 Range of cell densities obtainable in the different categories of fermentors used for animal cells

evolution of these fermentor designs and Figure 3.2 gives an indication of the range of cell densities achieved.

This section considers the benefits and the disadvantages of the various fermentation systems used and the design and performance limitations imposed on them by cellular physiology.

Supply of nutrients from the medium and supply of oxygen are the primary factors which govern growth and product yield of a given cell in culture. Oxygen must be continually supplied to cultures, and in small, low-density cultures, simple diffusion from the surface of the medium may be sufficient to supply demand. However, as will be discussed later, oxygen supply can become the primary limitation in scale-up and in achieving high cell densities.

The productivity of cultured cells is probably mainly limited by nutrient depletion and by accumulation of toxic metabolites which restrict the maximum cell density that can be achieved and the capacity of cells to synthesize product. Depletion of simple nutrients such as glucose and amino acids is rate-limiting for several cell culture systems, although, in other cases, the limiting nutrient may not be readily identifiable (Birch and Cartwright, 1982).

In principle, nutrient depletion can be readily resolved by addition of fresh medium in a so-called fed-batch process, and this approach has been very successfully applied to bioproduct generation (Tharakon and Chau, 1986). Fed-batch procedures can ultimately become ineffective as inhibitory waste products accumulate in the fermentor. This problem can be overcome by removal of spent medium from the fermentor as fresh medium is added, and this approach leads directly to perfusion systems in which fresh medium is continually passed through the fermentor to maintain near-optimum nutrient levels and to remove toxic metabolites. Product can also be continuously harvested from the medium stream. One technical problem associated with perfused fermentors is the necessity of retaining cells in the fermentor as medium passes through. With anchorage-dependent cells fixed to surfaces this poses no problem but with cells in suspension special provisions are necessary to prevent cells from washing out of the fermentor.

Batch versus perfusion cultures

One of the obvious advantages of perfusion systems in which cells are retained in the fermentor and medium continuously flows through the system, is the possibility of continuous product generation from the cells as long as these can be maintained under appropriate conditions of viability and productivity. In economic terms, several studies converge to indicate that, in biotechnological processes as in established chemical engineering processes, continuous operation is more efficient than batch processes (Schierer, Kanzler and Konopitzky, 1992; Runstandler, 1992). In animal cells, product release is frequently uncoupled from cell growth and occurs mainly when cultures are in the stationary phase. Long-term maintenance in this condition rather than continued production of new biomass is therefore the most effective route to efficient product generation. The key advantages in optimized perfusion culture operations are that the same biomass is employed for long periods before being discarded and that fermentation conditions can be held constant during the productive life of a given fermentation. This influences production costs as follows.

- Plant down time is reduced because the product release phase is prolonged.
- The unproductive growth phase is reduced as a proportion of the total process time.
- Product quality should be more consistent because conditions in the fermentor are held stable.
- Less personnel are required to service and supervise a continuously operating process.
- Cell densities and productivities can be achieved in a perfused system that cannot usually be achieved in batch culture.

Other advantages which are often cited may be more contentious. It is not clearly established that the specific productivity per cell (pg of product per hour per cell) is necessarily better in perfused systems. Neither is it clear that plant installations are less costly for perfused systems than for simpler batch processes.

Protagonists of batch processes point to the simplicity of the plant required, greater process flexibility (i.e. the capacity to produce different products in different production runs) and reduced quantities of product at risk since if a process failure occurs it is limited to a single batch

(Rhodes and Birch, 1988). In addition, scale-up parameters that are well understood in batch fermentors are much less well established for perfusion-based fermentations (Hu et al, 1991).

It is worth noting that, at the present date, the majority of really large-scale processes operating are still batch processes and that they frequently still use traditional technology such as roller-bottle culture for anchorage-dependent cells (Panina, 1985). Some completely new production operations also make use of this type of technology, and a recent particularly striking example of this has been in the production of erythropoietin (which is certainly the most profitable product derived from animal cells at present with sales exceeding a billion dollars annually) in roller bottles. In this case the primary motivation for this approach was the need to reduce development time to a minimum in order to reach the market before the competition. Production in roller bottles using existing technology is estimated to have saved 1–2 years over the time which would have been needed to create and validate a new production process and facility (Tso et al, 1991).

However, significant advances have been made in the application of "old" technology such as roller bottles, and an exciting development has been the application of robotics techniques permitting a much more reproducible and controllable use of this methodology which is otherwise labour intensive, tedious and vulnerable to contamination risks (Archer and Wood, 1992).

One other important consideration concerns the regulatory requirement for the clear definition of a batch of final product. In batchwise operation, the definition of a batch is self-evident. In continuous perfused systems, there is no natural break-point which describes a batch and it may be necessary to create such a break artificially, taking into account practical factors such as the genetic stability of the producer cells, stability of the product and appropriate volumes for down-stream processing. There appears to be no fundamental problem associated with this approach provided that it is validated and standardized.

Konopitzky, Kenzler and Windhab (1990) have proposed a system for monoclonal antibody production which combines the two approaches by using continuously operating perfused fermentors to provide the biomass to seed the final production fermentors which were operated in a fed-batch mode. The authors claim that this process combines the advantages of the continuous process with the regulatory clarity of an unambiguously defined batch of product.

Plug-flow versus stirred-tank fermentors

It can generally be assumed that, in stirred-tank fermentors, the concentration of cells, nutrients, waste products and the desired product are all homogeneously mixed in the culture. In batch operations nutrients become depleted during the course of a fermentation while product and waste metabolites accumulate in the fermentor. In stirred perfused fermentations, once the fermentor is fully populated, the concentration of these components tends to steady-state values. Medium flow is arranged so that none of the simple nutrients are limiting, and conditions of minimum medium flow capable of producing maximum product output can be determined and maintained. The maximum efficiency of medium use can thus be achieved.

The mixing characteristics in plug-flow systems are very different because medium is not homogeneously distributed through the fermentor and it is assumed that movement of materials through the fermentor are unidirectional with no back-mixing.

The mixing characteristics of these three fermentor configurations are illustrated in Figure 3.3.

An immediate consequence of plug-flow mixing characteristics are that the concentrations of nutrients and products experienced by the cells in plug-flow cultures vary not only with time, but also with the spatial position of cells in the fermentor. Gradients of nutrients, oxygen and cell products exist along the flow axis of plug-flow devices, and these represent a significant design limitation of fermentors of this type.

As will be discussed later, increased medium flow can be used to reduce gradients of metabolites, but this may introduce other problems such as increased shear and less efficient medium utilization. From an engineering viewpoint such possible heterogeneity renders scale-up more difficult. Adequate mathematical models for this type of reactor are not available and scale-up technology remains largely empirical (Hu et al, 1991). From a quality assurance point of view, the possibility that cells in different regions of the fermentor may be operating under different environmental conditions and perhaps at different doubling number because of differential growth rates can be difficult to reconcile with the need for total process control and consistent product generation.

Fermentor design for suspension cultures

Stirred-tank fermentors require sufficient agitation to maintain the cells in homogeneous suspension. Compared with microbial cells, animal cells are mechanically fragile and can easily be damaged by shear forces if excessive agitation rates are employed (Nevaril et al, 1968). The sensitivity of different cell lines to shear is reported to vary considerably. Shear forces can cause loss of productivity of cells in culture before any effect on cell viability can be observed (Fazekas de St Groth, 1983). To avoid shear damage, the lowest agitation rate consistent with efficient suspension is employed, and considerable effort has gone into the design of impeller systems able to give adequate mixing at low shear rate (Augenstein, Seniksey and Wang, 1971). Concern over possible shear

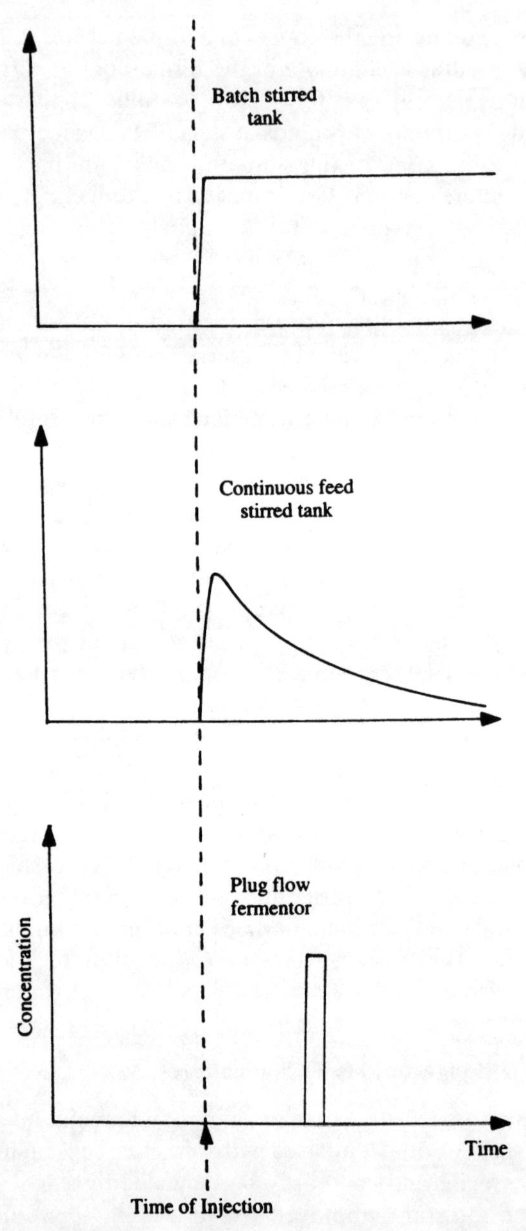

Figure 3.3 Mixing characteristics of different fermentor configurations; variation of concentration of an injected tracer at a given point in the fermentor after injection of the tracer at the time indicated.

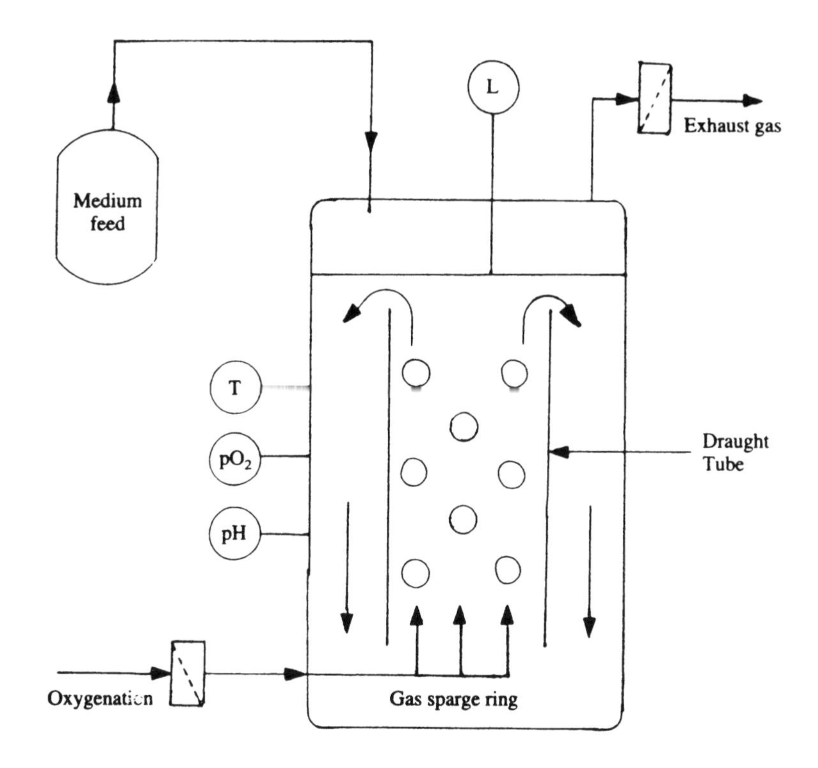

Figure 3.4 General arrangement of an airlift fermentor. Air from the sparge ring sets up a flow of medium in the draught tube and so serves to aerate the culture and act as a low shear agitator.

damage by impeller-based agitation systems led Katinger, Schierer and Kromer (1979) to evaluate the use of air-lift (also called bubble tube) fermentors for the culture of shear sensitive cells. In this system (Fig. 3.4), agitation is achieved without the use of a mechanical turbine by introducing air into the base of a draught tube within the fermentor. Bubbles rising in the draught tube induce overall medium flow as indicated in the diagram.

Hybridomas grown in serum-free medium have been reported to be rather sensitive to shear unless impeller tip speeds are kept below 0.19 m/s (Fazekas de St Groth, 1983). The inadequacy of oxygen transfer at these speeds led Birch, Boraston and Ward (1985) at Celltech to investigate the use of air-lift fermentors for the scale-up of monoclonal antibody production. Predictable scale-up of hybridoma culture up to 2000 litres was obtained, and air-lift fermentors form the basis of present industrial monoclonal antibody production at Celltech. Shear damage was reported not to be a problem and, even at the largest scale, the culture's oxygen demands were shown to be well within the oxygen transfer capacity of the fermentor. Another reported practical advantage of air-lift fermentation is its simplicity of design because it involves no moving parts or seals for rotating shafts such as are required with turbine-agitated vessels (Birch et al, 1985).

Airlift fermentors have also been used successfully for the large-scale generation of products from recombinant insect cells which have a high oxygen demand and which have also been reported to be unusually fragile (Shuler et al, 1990).

Oxygen consumption by cultured cells

For sustained cell viability it is essential that the dissolved oxygen concentration in the culture medium be maintained between critical limits. These limits vary considerably with cell type but usually lie between pO_2 levels of 30 to 200 mm Hg. Within these limits, cell productivity appears not to be very greatly affected by variation in oxygen tension (Boraston et al, 1983) but pO_2 levels above or below them result in powerful inhibition of cellular proliferation and product generation (Kilburn and Webb, 1968). These limits also determine the oxygen levels that are tolerable in the inlet stream and at the outlet of fermentors which are oxygenated via an external oxygenation system.

Oxygen demand rates also vary considerably between cell types from about 0.1 to 0.5 mmol/l/h at 10^6 cells/ml (Fleischaker and Sinskey, 1981). Because oxygen is poorly soluble in culture medium (about 0.2 mmol/l at 37°C), continuous input of oxygen is required. The oxygen transfer rate (OTR) from the gas phase into the culture medium in the fermentor must satisfy the oxygen demand. OTR depends on the area of the gas–liquid interface available for transfer, the difference in the oxygen concentration across this interface, and a mass transfer coefficient. In microbial fermentations with high oxygen demand oxygen transfer is achieved by sparging air into the culture, often at very high rates, and by using a high-speed turbine to break up the input air into very small bubbles to increase the surface available for transfer.

The sensitivity to shear forces of suspended animal cells precludes this form of vigorous aeration. In small-scale low-density cell cultures, the modest oxygen demands of animal cells mean that surface aeration can provide adequate oxygen transfer. However, as scale and cell density increase, much higher OTRs than can be achieved by surface aeration soon become necessary. Obtaining adequate OTR without producing shear-related cell damage remains one of the most critical process parameters requiring solution for the use of animal cells as efficient bioreactors on a production scale.

Direct sparging of air into suspension cultures of animal cells has been employed. However, due to the low agitation speeds employed because of shear sensitivity, bubbles cannot be fragmented by the impeller as in microbial fermentors. Hence the bubbles produced are large and their residence time in the fermentor is short so that a high sparge rate is necessary to obtain a useful OTR. Loss of productivity due to excessive foaming and to a direct detrimental effect of sparging on the cells has been reported (Telling and Radlett, 1970). The availability of microsparging systems able to generate bubbles below 0.5 μm in size may avoid some of these problems (Reiter et al, 1991).

The airlift fermentor uses injection of air into a draught tube to simultaneously aerate the fermentor and provide circulation of the medium. This system was successfully applied to suspension cultures of animal cells by Katinger et al, 1979). OTR values of 0.6–1.0 mmole/l/h were obtained, sufficient to satisfy the oxygen demand of most types of animal cells. Interestingly, Katinger showed that the gas flow per unit volume required for a given OTR decreases as reactor volume increases, indicating that oxygen supply would not be a problem in the further scale-up of airlift fermentors.

Aeration and shear damage

Because the problems of oxygenation and shear damage represent such an important limiting design parameter in animal cell fermentations, several groups have attempted to elucidate the mechanism of the observed cell damage. It has been demonstrated that cell damage in suspension is related to exposure of the cells to the small turbulent eddies in the micrometre size range (Kolmogorov eddies) which are supposed to dissipate most of the turbulent energy (McQueen, Meilhoc and Bailey, 1987). The Kolmogorov eddies produced become smaller as agitation speed increases (Croughan, Hamel and Wang, 1987) and maximum cell-disruptive effects are produced by eddies which approximate in size the cells which become subjected to violent pressure oscillations as they enter or leave the eddies.

Agents such as serum and methyl cellulose tend to protect cells against shear damage. One mechanism involved in this effect may be the increase in the size of Kolmogorov eddies produced by increasing the viscosity of the medium (Croughan, Sayrre and Wang, 1988) although other protective mechanisms have also been proposed (Goldblum et al, 1990).

It appears that there is a close relationship between mechanical cell damage and bubbles in the medium which are produced by vortex formation or cavitation at higher stirrer speeds. Kunas and Papoutsakis (1990 a, b) have shown that higher stirrer speeds can be tolerated by cells if the fermentor contains no head space so that bubble entrainment and breakup does not occur. However, at very high agitation rates cell damage does occur even in the absence of bubbles, suggesting that turbulent eddies in the liquid itself eventually cause cell disruption.

The role of bubbles in cell damage is of critical importance in fermentor design and operation. It is now widely accepted that the bursting of bubbles produces turbulent eddies which cause cell damage. This is supported by observations by Murhammer and Goochee (1990) showing that cell damage is produced as bubbles detach from the sparging apertures in airlift fermentors. It was proposed that this is due to turbulence produced in the liquid by liquid rushing to fill the space left by bubble detachment.

Nonionic detergents such as the polyol Pluronic F68 used as an additive, particularly to serum-free medium, have been widely shown to reduce cell damage in fermentors. It is currently postulated that one of the main effects of Pluronic is the stabilization of bubbles so that turbulence due to bubble bursting does not occur adjacent to cells. This effect is particularly marked at the culture surface where gas bubbles disengage from the medium. Surface foams produced in the presence of Pluronic F68 are more stable and drain more slowly than those produced in its absence, again reducing bubble collapse in the proximity of cells (Handa-Corrigan et al, 1991). However, these effects associated with bubble stability cannot be the only stabilizing influence exerted by Pluronic F68 because Cherry and Aloi (1991) have shown that the surfactant also stabilizes against cell damage produced by turbulent eddies in plug-flow systems in which no bubbles are present. Jobses, Marten and Tramper (1991) proposed that Pluronic F68 protects cells from shear by forming a stagnant film of fluid around the cells, thus avoiding their direct contact with turbulent eddies.

An alternative approach to avoiding cell damage during aeration is the use of oxygenation systems which avoid bubble formation completely. Several groups have used gas-permeable membranes either inside the fermentor (Thaler and Varnak, 1991) or assembled as a "lung" in an external circulation loop (Lehmann, Vorlop and Buntemeyer, 1988) to achieve this. The design of such systems is limited practically by the large surface areas that become necessary when fermentor volume increases. Liquid perfluorocarbons, which dissolve oxygen very efficiently, have also been used as oxygen transporters in fermentors (Hakamoto et al, 1987) although experience with this approach is still limited.

Perfused suspension systems

The fundamental engineering problem with perfused suspension cultures is that of retention of cells in the fermentor since animal cells sediment only slowly due to their small size and neutral density and are therefore easily washed out with the medium flow. Several approaches to cell retention have been incorporated into practical fermentor designs. These include cell-excluding filter devices (Himmelfarb et al, 1969), centrifugational cell retention (Hamamato, Ishimatu and Tokashiti, 1989; Tokashiki et al, 1990), membrane devices (Buntemeyer, Bodeker and Lehmann, 1987) and sedimentation chambers (Tokashiki et al, 1989; Tokashiki and Arai, 1991).

Filters and membrane retention devices can be located either in the fermentor or in external circulation loops whereas centrifugation invariably involves circulation external to the fermentor. Circulation through external loops may give rise to shear-induced cell damage in the circulating pumps and several centrifugation systems have also been reported to cause shear damage.

More recently however, continuous-flow centrifugation using a specially designed Westfalia CFA-01 separator has been shown to retain better than 99.99% of cells with no decrease in viability (Jäger, 1991). In this system cells were continually recycled to the fermentor without pelleting and in this way rotor fouling by the accumulation of cell aggregates was avoided.

With filtration devices, particularly those internal to the fermentor, filter fouling or blinding is the primary problem. The most widely applied solution to this problem has been the use of rotating filters (spin filters) to prevent cells and debris from attaching to the filter surface. However, shear sensitivity of cells in the proximity of the filter can still be troublesome (Himmelfarb et al, 1969).

The usual spin-filter configuration involves a cylindrical filter attached to a rotating shaft in the fermentor. This shaft can also serve to agitate the culture (via attached impellers) and provides a convenient route for withdrawal of the perfusate. A typical spin-filter arrangement is illustrated in Figure 3.5.

Various materials have been used to construct the spin filter, but the most common commercial designs use stainless steel mesh. The advantages of this material include good bio-compatibility, ease of sterilization with steam and ease of cleaning and re-use of the filters.

Not surprisingly, the major factors which govern the retention of cells in a spin-filter fermentation are the filter pore size, the spin speed and the perfusion rate (Himmelfarb et al, 1969; Varecka and Schierer, 1987) and these parameters can be accurately modelled using latex microspheres (Siegel, Fenge and Fraune, 1992). However, in real cell cultures the situation may be more complicated due to the specific properties of the cells and medium involved. Thus, when culturing CHO cells in a spin-filter fermentor, Fenge et al (1992) observed little effect on cell retention of either spin speed or perfusion rate over quite wide ranges. This was due, at least in part, to the tendency of the cells to form aggregates of varying sizes depending on the spin

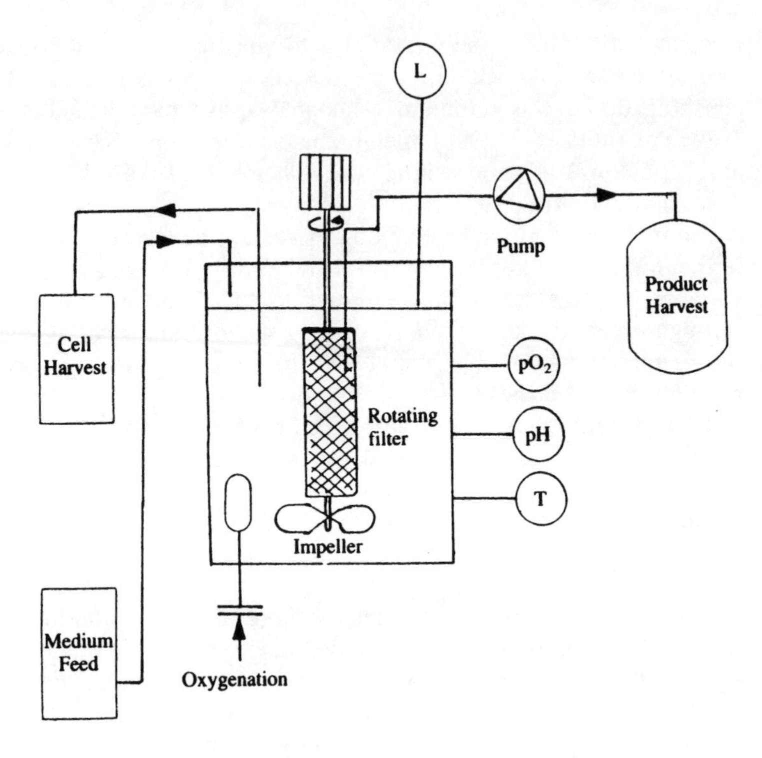

Figure 3.5 Typical spin-filter fermentor arrangement. In this system, an essentially cell-free product stream is withdrawn from inside the rotating filter. Adjustment of pH, pO_2 and temperature is performed in the fermentor.

speed. Lower spin speeds favoured aggregate formation but poorer retention of individual cells by the filter whereas high speeds improved the retention of individual cells but at the same time disrupted the aggregates.

The relationship between pore size and retention is also more complex than it might appear. Early studies employed spin filters with a pore size smaller than that of the average cell so that separation was based purely on a mechanical filtration effect, and filter blinding was a major problem. Much larger pore sizes have now been employed (Varecka and Schierer, 1987; Avgerinos et al, 1990), and it is widely accepted that in this situation cells are retained not by simple filtration but by their inability to cross the hydrodynamic barrier produced by the boundary layer generated adjacent to the rotating filter (Brenner, 1966; Favre and Thaler, 1992). Filters operating by this mechanism are far less readily blocked by cells and debris than filters with smaller pore size.

The configuration of the filter also affects retention performance and placing the rotating filter within a draught tube or adding a conical base to the filter have both been shown to affect performance.

The nature of the filter surface is also important, both in terms of the material used and texture (Esclade, Stephane and Peringer, 1991), which probably reflects both the stability of the boundary layer and the adhesiveness of the surface because adherence of protein from the medium and subsequent adherence of cells are important factors in blinding. Stainless steel appears to be generally satisfactory in this respect. The nature of the weave employed nevertheless influences fermentor performance (Jan, Emery and Al-Rubeai, 1992). The maintenance of high cell viability is also of critical importance because low viability tends to favour filter blinding (Siegel et al, 1992), possibly because of initial adhesion of cell debris and released proteins.

Aeration in spin-filter systems

Cell densities in excess of 10^7 cells/ml are readily obtained in spin-filter fermentors and oxygen demand soon becomes a problem. Direct sparging into the fermentor with use of antifoam greatly accelerates the blocking of the filter (Jan et al, 1992). Several approaches have been employed to avoid the problem, including sparging air into the inside of the rotating filter (i.e. away from the cell-containing space). Using this approach, an adequate OTR was reported by Thaler and Varnak (1991) for fermentors of 140 and 270 litres working volume. However, in other studies (Jan et al, 1992) it was not possible to obtain sufficient OTR by sparging high-density ($>10^7$ cells/ml) cultures inside the rotating filter.

The use of silicone membranes or tubes immersed in the fermentor has also been effective for bubble-free aeration of spin-filter systems (Thaler and Varnak, 1991; Jan et al, 1992).

Several practical industrial spin-filter fermentation units are currently available and are used for bioproduct generation. They combine several important advantages for industrial-scale production of animal cells derived products including

- high cell density (about 3×10^7 cells/ml).
- homogeneous fermentor conditions.
- direct control of cell culture conditions.
- proven scaleability (up to several hundred litres).

It should be noted, however, that scale-up is ultimately limited by the surface area of the filter that can be practically incorporated into the fermentor design. Use of high-performance filtration modules in a loop external to the main fermentor has been reported to overcome this problem to some extent (Sumeghy, 1992).

Fermentor design for immobilized cells

Anchorage-dependent cells

Some types of animal cells have an absolute requirement for attachment to a surface before they can proliferate or produce bioproducts. Scale-up of production of these cells, which had previously been restricted to multiplying the number of culture flasks or roller bottles used or by the use of multiplate vessels (Weiss and Schleicher, 1968), was revolutionized by the introduction of the microcarrier concept by van Wezel (1967). In this approach, cells are attached to the surface of near-neutral density microspheres in the size range of 100–400 μm (more typically 150–250 μm) after which the culture can be treated as a suspension culture system. A wide range of microcarrier particles have been developed, beginning with dextran beads and including polystyrene, cellulose, collagen and gelatin-based beads.

For anchorage-dependent cells, microcarrier culture provides the advantage of greatly increased surface area for cell growth and increasing the growth area per unit volume in the culture. Microcarrier cultures can be grown in conventional stirred fermentors, although some design modifications were needed to maintain the microcarrier in suspension without damaging the particle mechanically during agitation. This problem was largely resolved by the use of deep pitch marine type impellers revolving at low speed in vessels with hemispherical bottoms (van Wezel, 1967). Microcarrier cultures have also been grown in airlift fermentors and using perfused suspension culture techniques. Spin-filter systems are well adapted to microcarrier cultures because filters of large pore size can retain the culture very efficiently. Conventional (first generation) surface microcarriers of this type have found wide application in industry. However, they suffer from some important limitations:

- Because only the bead surface is used for cell growth, the beads represent a large 'dead volume' in the fermentor. This makes it difficult to obtain very high cell densities, and the large volume of beads required for higher-density culture presents engineering problems relative to the maintenance of an homogeneous suspension and aeration.
- Cell damage by abrasion between beads (bead to bead collision) can be significant.
- A large cell inoculum is required for efficient cell growth.
- The actual increase in biomass achieved during a fermentation can be modest because, with high seeding densities, the available bead surface soon becomes crowded.

Table 3.8 *Characteristics of Cultipsher® macroporous gelatin-based microcarrier beads*

Designation	Bead diameter μ	Pore Size μ	Density gm/ml	Representative Cell Types Grown
Cultispher G	220	10—20	1.04	CHO, BHK, L929, MDCK, HeLa
Cultispher GL	220	50—70	1.04	Endothelial cells, C127 Mouse mammary tumour
Cultispher GLD	510	50—100	up to 1.3	Various

- Processes for recovery of the cells and reseeding onto fresh beads are not very efficient.
- Single-use surface microcarriers are expensive and add substantially to the cost of a fermentation.

Macroporous microcarriers

One approach to avoiding the disadvantages of microcarriers listed is the so-called macroporous microcarrier. In this second generation of microcarriers, the beads are of open, reticulated structure and have surface pores which are large enough to permit entry of cells into the interior of the bead. Once inside the beads, the cells proliferate in an enclosed microenvironment where they are protected from mechanical stress due to liquid agitation and to collisions between beads. The surface area per unit volume available for cell growth is much higher than in surface microcarriers, and the minimum effective seeding density is lower (Almgren et al, 1991; Nikolai and Hu, 1992). Several forms of macroporous microcarriers are now available including those based on gelatin (Cultispher, Reiter et al, 1990; Almgren et al, 1991), polystyrene (Polyhipe, Lee et al, 1992) and polyethylene (IAM-carrier, Bluml et al, 1992b). It has been reported that up to 40% of the internal volume of cross-linked gelatin macroporous microcarriers can be occupied by cells (Nikolai and Hu, 1992). Macroporous microcarriers are available with varying pore size to suit different cell types, and their density can be varied by incorporation of various filler components (Table 3.8). This has led to the use of denser forms of the macroporous carriers in a fluidized-bed configuration rather than in true suspension. This approach will be discussed in a later section.

Another recent development has been the incorporation into macroporous carriers of cells such as hybridomas that are capable of growing in free suspension. The advantages of this are reported to include the protection of an optional microenvironment and protection against shear damage (Almgren et al, 1991).

Cell encapsulation

Another approach which deserves mention is the cultivation of cells in microcapsules. Microencapsulation was first proposed in the mid-eighties as a potentially useful method for hybridoma production because cells could be cultivated in a protective environment in a microporous capsule, and product recovery could be simplified by accumulation of the product in the capsule (Duff, 1985; Boag and Sutton, 1987). Microencapsulation has not been generally useful primarily because the microcapsules available (mainly alginate capsules) were not sufficiently robust for use in stirred fermentors and because animal cells could not tolerate the chemical treatment required to make the microcapsules more robust. Recent developments using polyethylenimine reinforcement of alginates have produced much stronger capsules compatible with high cell viability. Microencapsulation systems may, therefore, now become more widely used in bioproduct generation (Hsu and Chu, 1992).

Cell aggregates

Under certain culture conditions, anchorage-dependent cells may spontaneously aggregate to form spherical clumps which can then be propagated in suspension culture (Litwin, 1991). Cells grown in this way can be considered as a self-assembling microcarrier system and for some applications may represent a low-cost alternative means of generating biomass because no microcarrier particles are required and simple serum-free medium appears to be adequate to support the growth of aggregates of several commercially important cell types such as Vero, BHK and CHO (Litwin, 1991). Indeed it is interesting to note that for CHO cells, aggregate formation is favoured by serum deprivation (Jordan et al, 1992). The use of very small microcarrier beads in the 5–10 μm range also stimulate the formation of aggregates, presumably by providing nucleation sites for aggregation. Multilayers of cells (up to 20 layers thick) can be formed around the beads so that very high cell densities can be obtained using very low concentrations of beads (Goetghebeur and Hu, 1991).

A possible limitation on the use of aggregate cultures is their apparent sensitivity to shear damage. Jordan et al (1992) showed that serum protected aggregated CHO cells against death by shear but that the protection was actually due to a reduction in the size of aggregates arising in the presence of serum and to the fact that smaller aggregates were less shear sensitive.

However, production of bioproducts from aggregate cultures may also be difficult because the variation in aggregate size and the occurrence of necrotic regions in the centre of large aggregates produce an inhomogeneity of the culture and subsequent difficulties in controlling and monitoring cell growth and production conditions. Several recent studies have investigated in detail the parameters which govern aggregate size and viability in suspended aggregate cultures (Renner, Jordon and Eppenberger, 1993).

Aggregate cultures have been operated in both batch and perfused suspension systems (Murata, Eto and Shibai, 1988) and seem particularly well adapted to spin-filter fermentors (Avgerinos et al, 1990). The successful culture of cell aggregates in high-density fluidized-bed fermentors has also been reported (Reiter et al, 1992b). Aggregates of hepatocytes formed in suspension culture have also been immobilized on membranes and shown to retain the characteristic synthetic capacity of differentiated hepatocytes over longer periods than the same aggregates maintained in suspension (Sakai, Furukawa and Suzuki, 1992).

Plug-flow fermentor systems

A variety of cell culture devices have been developed in which the cells are maintained immobilized while the medium stream flows past them. Critical design parameters for such systems include assuring the homogeneity of medium flow and cell growth conditions radially across the cell bed and avoiding significant gradients of nutrients, oxygen and toxic products axially. Many ingenious technical solutions have been applied to this problem and some of these have been incorporated into practical production scale fermentor designs.

Because the cells remain immobilized in the reactor bed of such fermentors, the medium can be oxygenated and otherwise conditioned in an external circulation loop, thus avoiding the problem of cell damage during aeration. Figure 3.6 shows the general arrangement for a typical fixed-bed fermentor.

Fixed-bed fermentors

Glass beads were used in a packed-bed fermentor in early attempts to scale-up anchorage-dependent cell cultures (Earle, Bryant and Schilling, 1954), and this approach was later developed into a practical system for virus production by Spier and Whiteside (1976). Many other types of support have since been studied including a variety of shapes of ceramic particles. These include ceramic cylinders and spirals of the type mass-produced for use in condenser and catalytic columns in the chemical industry.

Problems associated with fixed-bed reactors of this type include channelling of liquid flow giving rise to inhomogeneities in the bed and the

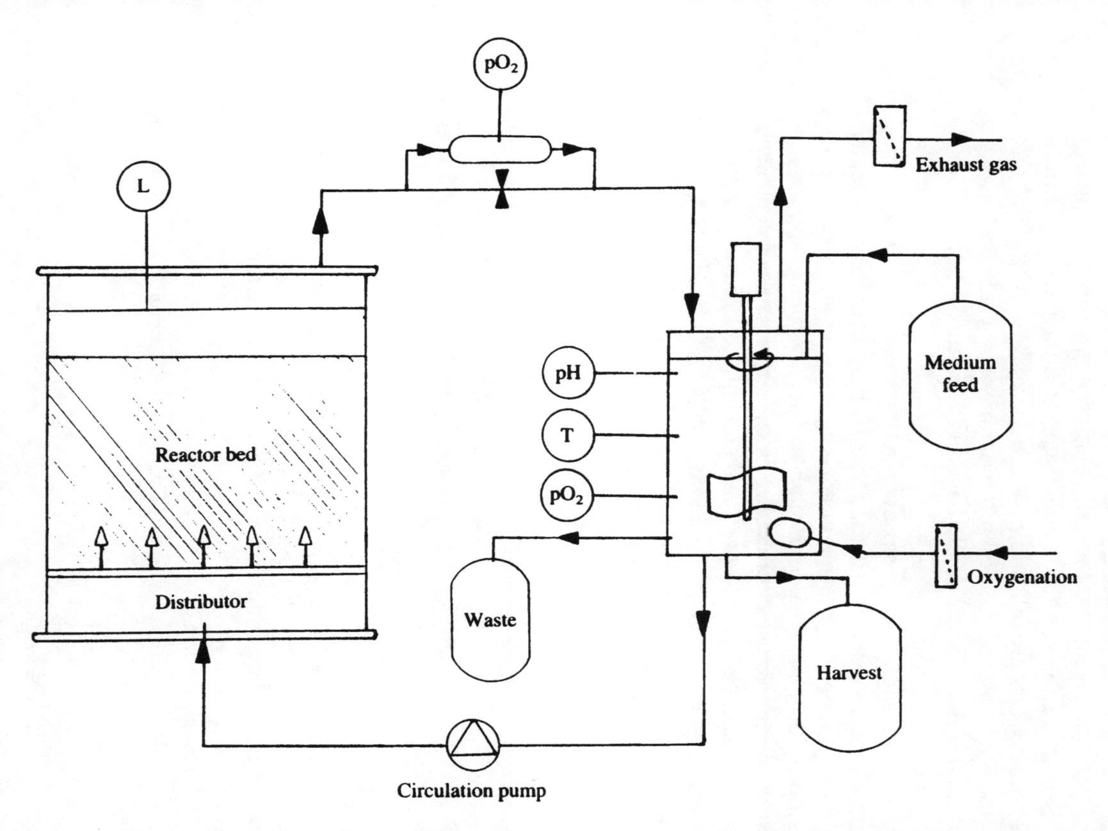

Figure 3.6 Typical fixed-bed reactor arrangement. Medium is pumped through the bed of the reactor by a circulation pump. Medium is conditioned (pH, pO_2 and temperature) in a separate stirred vessel. Product can be continuously withdrawn from the circulating medium stream.

formation of gradients along the bed length. Bead size and medium-circulation rate both influence these effects critically. Bead sizes of 2–5mm have usually been employed. At smaller bead size, bridging of the internal spaces by cells may result in channelling and flow-pattern heterogeneity. The problem of axial gradients has been reportedly overcome in recent studies by Bliem et al (1991) by the use of several stacking baskets of packing matrix in a single large fermentor.

As in the microcarrier situation previously discussed, the availability of microporous beads has provided a way of significantly increasing the cell density in packed-bed fermentors. In cases using macroporous beads, in which a large proportion of the bead volume is colonized by cells, smaller bead sizes have to be used to achieve adequate mass transfer characteristics.

In addition to macroporous glass beads (Looby and Griffiths, 1990) a number of other open-structure particles and matrices such as magnesium aluminate beads (Park and Stephanopoulos, 1993), fused alumina foams (Lee et al, 1991), polyurethane foam (Matsushita et al, 1991) and fabrics as either packed discs (Wang et al, 1992) or fixed sheets (Robert, Côté and Archambault, 1992) have all been employed in fixed-bed reactors.

Porous ceramic material has also been formed into a solid cartridge for animal cell growth and was commercialized in this form by K C Biologicals as the 'Opticell' system (Bogner, Pugh and Lydersen, 1983; Berg and Bodecker, 1988). The Opticell reactor is constructed as a ceramic cylinder with 1 mm^2 channels or pores running lengthwise through the cartridge. A tendency to blockage of the pores has limited the utility of this system.

Although very useful results have been obtained using fixed-bed fermentors, scale-up potential is limited by the problem of concentration gradients in the larger bed sizes. One possible solution is the construction of multiple beds in a single fermentor as already mentioned (Bliem et al, 1991) or the operation of multiple separate beds in parallel. The use of radial-flow systems also deserves evaluation. However, the most practical approach to assuring homogeneity of the bed is the use of fluidized-bed technology.

Fluidized-bed fermentors

In fluidized-bed systems, particles are maintained in very dense suspension and continually mixed by upward medium flow through the bed. Mass transfer characteristics are therefore much improved compared with fixed-bed systems. Because the reactor bed is constantly in motion, problems such as channelling which are seen with fixed-bed systems are avoided. This permits the use of smaller carrier particles in the bed.

Table 3.9 *Commercially available macroporous carriers for fluidized-bed operation*

Carrier	Material	Diameter μm	Density g/ml	Pore size μm	Void Volume %	Manufacturer
Microsphere	Collagen	500-600	1.6	20-40	75	Verax Corporation
Siran	Glass	300-5000	—	10-400	60	Schott Glaswerke
Informatrix	Collagen - glycose aminoglycan	500	—	40	99.5	Biomet Corporation
Cultispher GLD	Gelatin	510	1.3	50-100	80	Percell Biolytica
IAM-Carrier	Polyethylene	1500-2000	1.3	100-300	75	IAM Vienna

Smaller particles are desirable for the high-density culture of animal cells in macroporous supports so that cells growing inside the particles can be efficiently supplied with nutrients and oxygen. These characteristics have led to intensive investigation of fluidized-bed fermentors for the large-scale generation of products from cultured animal cells (Runstandler and Cernek, 1987; Runstandler and Cernek, 1988; Looby and Griffiths, 1990). In fluidized-bed reactors, cells do not grow on the exterior surface of particles where they would be dislodged by interparticle abrasion, but, as with the macroporous microcarriers, they colonize the internal surfaces or spaces of the particle where they proliferate in a protected microenvironment.

Cell carriers used in fluidized beds include purpose-designed macroporous supports such as Siran[R] glass beads produced by Schott and collagen Microspheres[R] produced by Verax. Other particles have been derived from macroporous microcarriers that have been engineered for fluidized-bed operation by the inclusion of materials to increase their density. Examples of this include Cultispher GD[R] and GLD[R] from Percell Biolytica (weighted with titanium oxides) and IAM-carrier polyethylene beads (weighted with silica). Characteristics of several macroporous supports for fluidized-bed use are given in Table 3.9).

Fluidized-bed fermentor design

Most fluidized-bed reactors conform to the same general configuration as the fixed-bed reactor shown in Figure 3.6, with medium flowing upwards through the matrix of carrier particles. Particles are designed to be light enough to be fluidized effectively by practical medium-flow rates but of sufficient density to be retained in the chamber by sedimentation. The volume of the fluidized-bed is typically 1.2–1.5 times that of the same carriers at rest.

Recently an integrated design has been developed for a fluidized-bed fermentor in which no external medium circulation loop is required because all the necessary elements for agitation and oxygenation are included in a single vessel (Reiter et al, 1991). In this system the bed of macroporous polyethylene beads was fluidized by a low-shear axial impeller acting in a draught tube. Oxygenation was by a 0.5 μm-pore sparging system in the fermentor (Figure 3.7). It was calculated that, with pure oxygen, 1 cm^2 of microsparger surface was sufficient to supply the oxygen demands of 10^{12} cells. The required OTR was achievable at low gas-flow rate, and no antifoam was necessary. In this system, up to 75% of the fermentor volume could be filled with cell-containing macroporous beads (Reiter et al, 1992a).

An alternative approach to the supply of sufficient oxygen to very high cell density cultures in fluidized beds has been to include oxygen-permeable tubes arranged vertically in the fermentor. This tends to prevent the formation of oxygen gradients along the flow axis which can limit the capacity for scale-up of fluidized-bed fermentors (Hambach et al, 1992).

Hollow-fibre cell culture systems

Hollow-fibre systems were initially conceived as artificial tissues comprising a cell mass growing in a chamber fed with nutrients via a series of porous capillary tubes through which medium is circulated (Krazek et al, 1972).

Generally, hollow-fibre devices are adapted forms of the widely used hollow-fibre ultrafiltration, microfiltration or dialysis modules and consist of a bundle of capillary tubes of diameter of the order of 3–400 μm and of porosity which can be varied to suit the application. The capillaries are usually arranged to act as microporous filters or as ultrafiltration membranes with a cut off of 10–100,000 molecular weight. On a small scale, disposable hollow-fibre cartridges of the type used for kidney dialysis are frequently used. Steam-sterilizable cartridges are now available for production-scale operations. Cells are seeded into the space surrounding the capillaries, and medium is perfused through the capillaries (Figure 3.8). Product accumulating in the extracapillary cell compartment can be periodically or continually withdrawn from the fermentor.

Cell growth in the extracapillary space offers several practical advantages:

- Cells immobilized in the fermentor can be grown to very high densities by perfusion of medium through the capillaries.

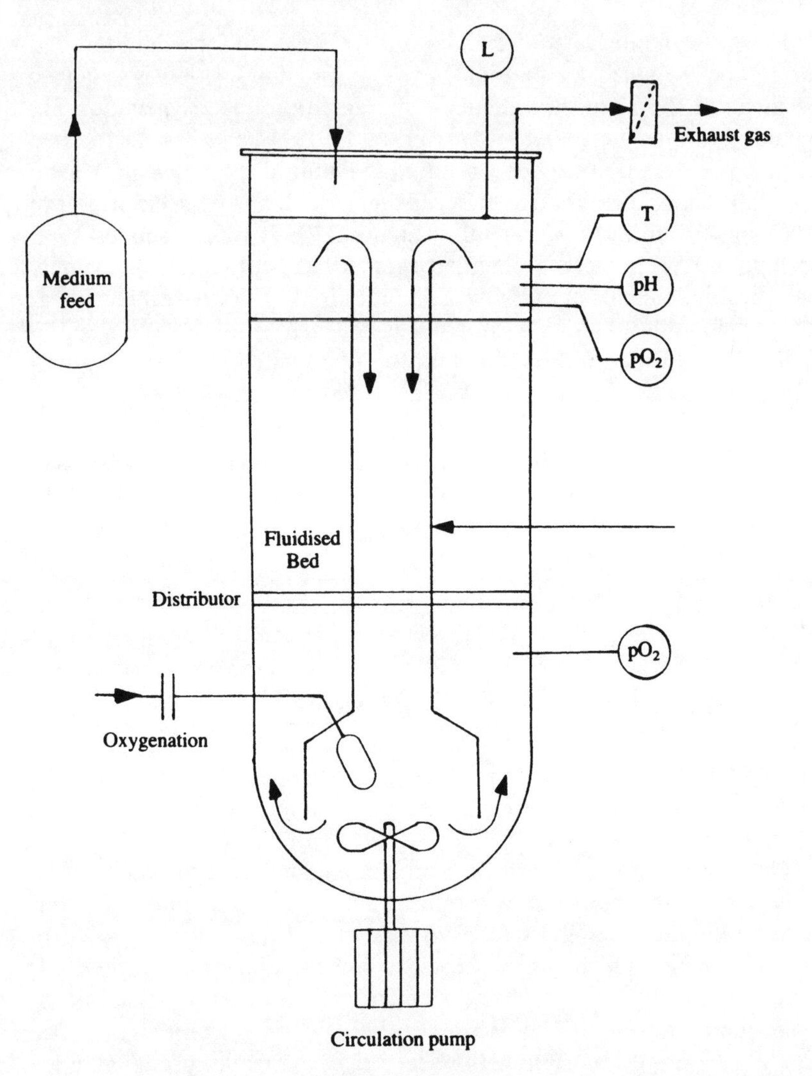

Figure 3.7 Integrated fluidized-bed fermentor general arrangement (after Reiter et al, 1992a). Medium is pumped up through the fluidized bed after oxygenation by a microsparging system located in the draught tube.

- Oxygenation, which occurs through the capillary wall, does not involve contact of the cells with bubbles.
- Cells are not subjected to shear forces.
- Serum or other macromolecular medium supplements can be used very economically because they need only be present in the low-volume extracapillary space; often no exogenous macromolecules are required because once the cell culture

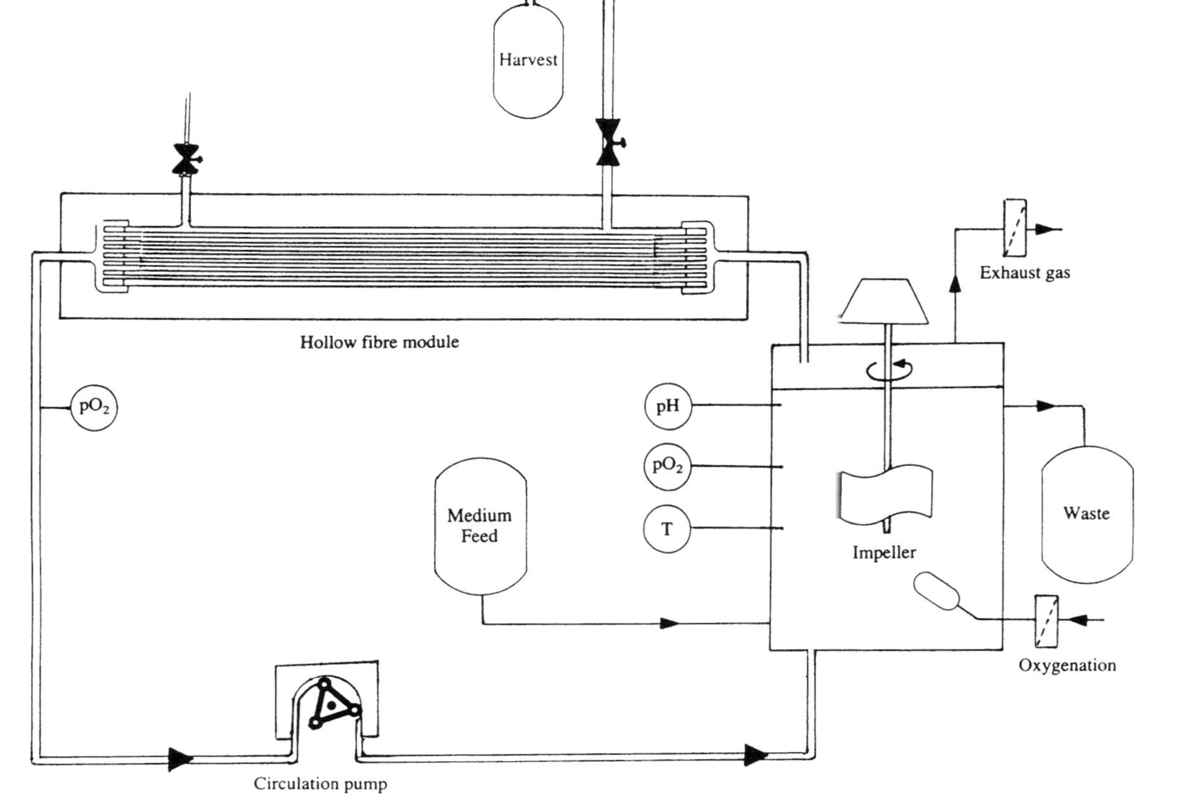

Figure 3.8 Hollow-fibre fermentor, general arrangement. Medium is circulated through the capillaries and is conditioned (pH, pO_2 and temperature) in a separate vessel. Product is retained in the extracapillary space with the cells and can be withdrawn either continuously or periodically through the sample port as illustrated.

is fully established, products secreted by the cells themselves are sufficient to maintain viability.

- High concentrations of product can be accumulated in the extracellular space; if no exogenous protein is added into this space, product can be accumulated at high purity, thus simplifying downstream processing requirements.

Disadvantages of hollow-fibre fermentors

Disadvantages of hollow-fibre fermentors include the extreme heterogeneity of the system which allows the formation of both axial and radial gradients of nutrients and metabolites. The main biochemical engineering design parameters governing such gradients are fibre diameter, fibre packing density, fibre path length and medium recycle rate. In practice, oxygen transfer appears to be a primary limiting factor because of the low solubility of oxygen in aqueous medium, and pH, lactate and ammonium gradients have been reported to be much less important (Piret and Cooney, 1991).

In a careful study of radial diffusion from porous fibres, Sandonini and Di Biascio (1992) examined the pattern of growth of cells immobilized in agarose around single fibres perfused with medium. Cell growth diminished radially from the fibre at a rate which depended entirely on the oxygen transfer rate. It was concluded that cells become anoxic at practical perfusion rates if the between-fibre spacing exceeded about 220 μm. This is within practical design limits, although increasing the number of fibres in a given section becomes limited by technical difficulties in effectively sealing the fibres.

Axial oxygen concentration gradients remain a significant problem. Higher medium flow rates can be employed to reduce the oxygen gradient, but this creates other problems due to the induction of a secondary flow, known as Starling flow, in the extracapillary space. This results in the formation of axial gradients of both cells and macromolecules in the extracapillary space, again producing heterogeneities in the fermentor (Piret and Cooney, 1991).

One approach to combatting this problem has been the introduction of a second set of hydrophobic capillaries specifically for gaseous exchange (Cousins, Gergen and Gerner, 1992). This system was claimed to reduced axial oxygen gradients and to simplify the fermentor because the need for an oxygenator in the medium circulation loop was eliminated and medium circulation could accordingly be much slower. The problem of radial gradients developing due to altered fibre packing was not addressed in this study.

It has been shown that Starling flow-induced gradient can result in decreased reactor productivity and that this effect can be reduced by periodically reversing the direction of medium flow in the capillaries.

Table 3.10 *Elements for and against the use of hollow-fibre cell culture systems*

Advantages	Disadvantages
Elevated cell density (ca 10^8/ml)	Inaccessability of the cells for inspection or for monitoring their viability or concentration
Physical separation of the cells from medium flow therefore cells are protected from shear forces even at high flow rates	Probable heterogeneity of the cellular environment due to axial and radial concentration gradients
Economy of macromolecular nutrients or growth factors which are separated from the main medium flow	Possibility of pockets of cells in locally adverse culture conditions
Ease of recovery of macromolecular products	Induction of concentration gradients of macromolecules due to Starling flow
Scale-up for research use is well established	Recovery of cells difficult
	Capacity for scale-up for industrial use has not been demonstrated

The polarization of macromolecule concentration induced by Starling flow has been used by some workers to increase harvest by recovering product from the end of the fermentor where macromolecules are concentrated (Piret and Cooney, 1990).

Another difficulty experienced with hollow-fibre reactors is that, in common with other high-density, immobilized cell configurations, direct monitoring of the production cells is not possible. Nuclear magnetic resonance techniques have recently been applied to this problem, and these studies tend to confirm that channelling and considerable heterogeneity in cell growth can occur at different points in the extracellular space due to the nonuniformity of medium flow in the capillaries and lack of mixing in the extracapillary space. Necrotic pockets of cells have been observed, and these could adversely affect product quality because of lytic enzyme release. Recent hollow-fibre fermentor designs have emphasized flow patterns which promote mixing in the extracapillary space to minimize the formation of local anoxic pockets. Periodic flow reversal in the capillaries appears to help in this respect as does pulsed alteration in the pressure differential between the two compartments. Another practical disadvantage of hollow-fibre fermentors may be the impossibility of reusing most types of hollow-fibre devices. The general points for and against hollow-fibre fermentors for production from animal cells are summarized in Table 3.10.

New developments and scale-up in hollow-fibre fermentors

Other new developments in hollow-fibre fermentor design include systems with separate feed flows for low molecular weight and macromo-

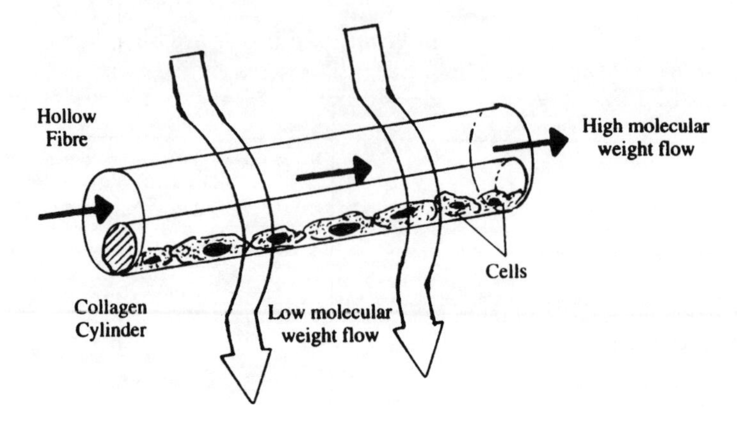

Figure 3.9 Concentric hollow-fibre culture system. In this system, cells are seeded into a collagen solution which is then allowed to gel. The gel contracts from the capillary wall with the cells entrapped in it, creating a separate flow channel within the capillary. This channel can conveniently be used for adding or removing macromolecules while leaving the cells in place in the capillary. This permits the use of separate flow streams for macromolecules and for low molecular weight nutrients.

lecular nutrients by use of concentric fibres with different molecular weight cutoffs. This has been achieved directly by using two artificial fibres and also by making use of collagen gel. In this system, cells are seeded into the lumen of fibres in a collagen solution which subsequently forms a gel and contracts away from the walls of the capillary, thus forming a new flow channel (Scholz and Hu, 1992). The cells are retained, entrapped in a cylinder of gel (Figure 3.9). Both of these systems permit the use of two perfusion streams, one luminal to remove or replenish macromolecules and the other extracapillary for low molecular weight nutrient supply as in the simple hollow-fibre configuration. The two streams can be operated at different flow rates to suit product generation rate and nutrient demands. This approach has recently been successfully applied to immobilized hepatocytes and may form the basis of an artificial liver device (Nyberg et al, 1993).

Several fully instrumented hollow-fibre cell culture devices are available and are used commercially for the production of proteins, particulary monoclonal antibodies, on the gram scale. As discussed earlier, further scale-up is limited practically by difficulties in the manufacture of wider bundles of fibres and axially by the oxygen gradient problem. The most practical solution, advocated by the manufacturers, is the use of several hollow-fibre modules connected in parallel. Although this approach appears feasible, detailed information on the successful scale-up of protein production to an industrial scale using hollow-fibre fermentor technology is not available at present.

Table 3.11 *Comparison of stirred-tank fermentors and immobilized cell systems for the cultivation of animal cells*

Stirred Tank Fermentor	**Immobilized Cell System**
Production classically by batch (may be perfused)	Capable of continuous operation over several months
Cells relatively dilute (2x106/ml) ($2x10^7$ with perfusion)	Cells may be very concentrated ($2x10^8$ /ml)
Product concentration low	Product concentration high
Solid/liquid separation needed	Not needed
Scale-up well understood	Scale-up may be a problem
Process control relatively simple (homogenous system)	Control may be complicated
Cell recovery easy	Cell recovery difficult

Table 3.12 *General characteristics of the fermentor configurations available for animal cells*

Fermentor configuration	**Cell density (millions/ml)**	**Complexity of operation**	**Capable of scale-up**	**Homogenous**
Agitated tank fermentor (batch or fed batch)	1-4	Simple	Yes	Yes
Perfused tank fermentors ■ Spin Filters ■ Dialysis Systems ■ Centrifugation	30-70	Intermediate	Yes	Yes
High Density Systems				
■ Fluidized bed (e.g. Verax)	>100	Complex	Probably	Yes
■ Hollow fibres (e.g. Endotronics, Acusyst)	>100	Complex	Unknown	No
■ Ceramic matrix (e.g. Opticell)	>100	Complex	Unknown	No

Overview

A variety of fermentor configurations exist that have been employed for the culture of animal cells. Each has specific features, discussed in this chapter, which influence their suitability for biopharmaceutical manufacture in terms of scale-up capacity, regulatory acceptability and effects on downstream processing. These key characteristics are summarized in Tables 3.11 and 3.12.

Process control

High-density cell culture presents another challenge to the process development engineer at the level of process control. As mentioned, conditions in classical, low-density cell cultures change slowly and satisfactory results can be obtained with minimal control. The same is not true at high cell densities in which rapidly fatal results may follow if oxygen tension or pH moves outside the rather narrow limits required for cell viability. Because they function effectively only in a narrow band of acceptable fermentor conditions, animal cells can be more vulnerable to minor perturbations than most microbial fermentations. Thus, failure of a circulating pump or oxygen electrode cannot be tolerated in high-density cultures. The tendency to remove serum and other protective proteins from the culture medium to facilitate downstream processing tends to render the cells even more susceptible to perturbations in the fermentor environment.

At the same time the invested and potential value of the fermentor increases as cell numbers and cell density increase so sensitive and reliable systems of control and alarms for deviation from preset process limits are required. Furthermore, as the scale of operation increases, manually operated plant becomes labour intensive and hence more vulnerable to operator error. Automated plant control offers numerous advantages in terms of reproducible, error-free operation and rapid and appropriate responses to changes in fermentor environment. However, it is difficult to automate and control systems that cannot be effectively monitored, and the development of continuous on-line sensors is an area where major efforts are currently being made.

The fully optimized use of animal cells as bioreactors ideally requires measurement of the parameters listed in Table 3.13 and an understanding of their impact on productivity.

In-fermentor probes

Measurements of pH inside the fermentor are now practical and reliable following the development of sterilizable pH electrodes with robust glass membranes. Drift in pH electrode performance can occur, but this is minimized by the use of probes which incorporate a fluid bridge to avoid fouling of the reference electrode junction caused by precipitation of medium components. Control to within ± 0.2 pH units may be necessary to maintain optimum pH performance.

Reliable oxygen electrodes are also available, although electrode fouling during long fermentation runs is still a problem. Modern oxygen electrodes show only limited drift (around 0.5–1% per day) although-

Table 3.13 *Parameters which would ideally be monitored in animal cell bioreactors*

Parameter	Availability "in fermentor"	Availability of specific sensors
Temperature Dissolved oxygen pH Redox potential	Yes	Efficient electrodes and probes are available
Cell viability	No	Released enzymes Fluorescence
Cell concentration	No	Indirect measures only
Depletion of critical nutrients	No	Specific biosensors
Accumulation of toxic products	No	Specific biosensors
Protein concentration Product concentration	No	Automated immunoassay, near on-line HPLC, CE, MS, etc

steam sterilization may cause much more rapid modification of electrode response. Periodic recalibration is therefore necessary and designs which provide filtered air vents into the sensor chamber are used to permit calibration in situ.

In addition to their main use for controlling of pO_2 within the limits required by the culture, oxygen electrodes permit the determination of the oxygen uptake rate of the culture and, hence, can give an indirect measure of cell numbers. However, as already discussed, oxygen uptake rates can vary considerably, depending on culture conditions (Boraston et al, 1984; Miller et al, 1987), and its use to estimate cell concentration requires careful validation in the specific fermentation conditions used.

In homogenous stirred fermentors, oxygen consumption rate can be determined directly from the decline of pO_2 with time when oxygenation is briefly stopped, and in plug-flow systems it can be calculated from a knowledge of pO_2 in the inlet and outlet medium streams and the flow rate.

Redox potential can also give useful on-line information about the state of the cells in culture (Griffiths, 1984). Redox electrodes are available which are not subject to fouling after prolonged use in the fermentor. The redox potential of medium in a cell culture decreases during the fermentation as a result of the reducing activity of cells and the accumulation of thiols (Glacken, Adema and Sinskey, 1988). If pH, pO_2 and temperature in the fermentor are all maintained constant, redox potential provides a meaningful measure of the number of viable cells in the culture. Since culture medium has only a very low redox buffering capacity, the measurement of redox potential can be a very sensitive monitor of changes occurring in cell number or physiological state (Hwang and Sinskey, 1991). The measured redox potential can be affected by several factors in the fermentor, including cell number, cell viability, substrates available for oxidation, pO_2 and availability of metabolic cofactors. Redox potential is therefore useful for monitoring fermentor kinetics to detect deviations from an established fermentor profile but can be difficult to interpret in absolute terms.

Intracellular fluorescence

Another approach to measuring the effective intracellular redox state of cells in the fermentor is the use of fluorescence to measure intracellular NAD(P)H concentration (Armiger et al, 1986). In this system, a fibre optic fluorescence probe is introduced into the culture to measure the fluorescence at 460 nm after excitation at 360 nm. Fluorosensors of this sort have been used to monitor changes in pO_2, which is reflected by changes in intracellular redox potential and hence changes of intracellular NAD(P)H concentration. This signal, coupled to the fermentor's oxygenation system, can then be used to control dissolved oxygen (Scheper, 1990). If the culture redox potential is maintained constant, fluorescence measurements of this sort can, in principle, be used to monitor cell numbers.

Fluorescence probes are subject to several practical limitations, including the possibility of the fluorescence signal being altered through fouling of the window through which cells are observed and the fact that only a very small proportion of the cell population can be observed. This observed part of the population may not be representative of the whole culture, particularly in heterogeneous systems such as hollow-fibre fermentors. Once again, another complication in interpreting the signals obtained is that they can be affected by many environmental factors including variation in pO_2, availability of substrates for oxygenation and the presence of toxic agents which may modify the efficiency of the oxidative process. All this information can be very useful in understanding the physiology of the culture but may be too complex to be fully useful in basic process control.

Biosensor technology applied to fermentation control

Biosensors are devices by which a biological sensing element such as an enzyme or an antibody is coupled to a transducing system capable of producing an electrical signal. The output of biosensors can be produced directly using amperometric or potentiometric systems such as enzyme electrodes or by measuring fluorescent changes in the so-called 'optrode' configuration (Scheper, 1990). Devices such as the light addressable potentiometric sensor (LAPS) developed by the Molecular Device Corporation translate chemical reactions occurring on the surface of a silicon chip directly into a potential which can be measured by its modulating effect on light illuminating the chip. Specific biosensors have been developed for measurement of several analytes important in cell culture including lactate, glucose, pyruvate and so on. Accurate, near on-line measurement of these metabolites can be very useful for piloting a fermentation and for rapid detection of deviation from a standard fermentation profile.

Glucose uptake and lactate production have been proposed as parameters that can be used as a measure of cell number. However, both of these parameters vary greatly with changes in the culture environment and, though useful for following global changes in the culture, cannot be reliably related to cell number.

So far, only the first generation biosensors, the amperometric enzyme electrodes and optrodes are generally available. Concentration of the relevant analyte is usually determined by measuring the activity of an enzyme which uses the analyte as a substrate and whose activity can be linked, through a second enzyme if necessary, to the oxidation or reduction of NAD, NADP or some other factor which can then be measured electrically or fluorimetrically. Extreme specificity of measurement of a given analyte can therefore be obtained by using a highly specific enzyme.

A recent development of particular interest for the monitoring of cultures of animal cells makes use of an enzyme electrode incorporating glutaminase and glutamate oxidase for the direct measurement of glutamine by flow-injection analysis. However, a complication of this system is the need to remove endogenous glutamate from the medium using an ion-exchange cartridge before reliable glutamine measurements can be obtained (Cattaneo and Luong, 1993).

A fundamental limitation of these biosensors is that they require an active enzyme (or possibly antibody) for signal generation and therefore they cannot be sterilized or included in the fermentor. Analysis must therefore be performed on samples withdrawn from the fermentor. This can be achieved by continuous withdrawal of medium through a microfiltration or ultrafiltration module, but the utility of the system very rapidly becomes limited by membrane fouling. If membranes are not used, sample withdrawal involves the risk of breaching fermentor sterility. There is thus a real need for the development of safe and effective continuous sampling devices for animal cell fermentors. An ideal sampler should remove a sample of medium that is representative, is sterile and which does not alter between removal from the fermentor and arrival at the point of analysis. Table 3.14 summarizes the main characteristics of biosensors and their utility for monitoring of fermentation runs.

Despite their extreme specificity, most types of biosensor cannot be

Table 3.14 *Advantages and disadvantages of biosensors (essentially enzyme electrodes and optrodes) for monitoring cell culture growth, viability and productivity*

Advantages	Disadvantages
Short analysis time High specificity Can be used in near on-line assays	Fragile: - cannot be sterilized therefore cannot be placed in the fermentor - may become poisoned or inactivated Efficiency may be limited by rapid membrane fouling, etc. Calibration and validation may be complicated. Precision and reproducibility may be inferior to conventional sensors.

readily adapted to measure the release of required products from cultured animal cells directly since these rarely have activities measurable in enzymatic terms. However, the LAPS devices can directly detect interaction such as the formation of antibody–antigen complexes or the interaction between receptors and effecters and could therefore, in principle, be adapted for direct product measurement. Measurement of product will be discussed briefly in a later section.

Measurement of cell viability

The measurement of cell viability when cells cannot be analyzed microscopically or by vital staining represents a significant technical problem. In systems conceived for the commercial manufacture of bioproducts, any cell which becomes incapable of producing product is effectively dead. The ideal functional indicator of cell death would therefore be the shut-off of product formation by the cell.

Unfortunately, this cannot usually be practically measured. Indirect indicators such as oxygen or glucose consumption give some idea of the viable cell count but, as discussed earlier, these parameters can be affected by many environmental factors and may be difficult to interpret.

Release of intracellular enzymes

Release from the cells of enzymes which are normally localized inside the cells has been employed as a useful marker of cell damage and death. The enzyme-release approach was initially applied to a microplate assay for virus induced cytopathic effect (Cartwright and Swain, 1974) using lactate dehydrogenase and glutamate-oxalacetate transaminase.

More recently, release of lactate dehydrogenase (LDH) has been proposed as a reliable indicator of cell damage in hybridoma cultures (Peterson, McIntire and Papoutsakis, 1988). Adenylate kinase release has also been employed for this purpose (Merten et al, 1992).

LDH release has been the most fully studied and is probably the most satisfactory cell death indicator for most cell types. Marc et al (1991) have shown that, for a given cell type under given culture conditions, LDH content per cell remains quite constant. LDH is released from damaged cells before cell lysis occurs and so LDH release provides a useful early indicator of cell damage and loss of productive capacity. However, the LDH content of cells is affected by several culture parameters, including pO_2, carbon source type and availability and pH. It also varies widely with different cell types so LDH release has to be calibrated for a given cell type under specified culture conditions.

One key question is that of the stability of the released enzyme in culture medium. Again this can vary with cell type, but most reports indicate that LDH is stable in medium at 37°C for many hours.

Use of Pluronic F68, which is frequently added to medium as an antifoam agent or to stabilize against shear stress, may invalidate estimates of cell death based on LDH release since Pluronic interferes with the mechanism of enzyme release and therefore can lead to underestimation of cell death (Gardner, Gainer and Kirwan, 1990).

Recently, Bowes and Bentley (1993) have shown that LDH can be used to reliably estimate both viable and total cell counts in a CHO population. This was achieved by measuring LDH activity in the culture supernatant after cell removal (as an indicator of the number of dead cells present or lysed in the culture) and the total quantity of LDH released from a sample of the culture when *all* cells in the sample were lysed by detergent. Since LDH content per cell remains constant under given culture conditions, it is possible to calculate total cell numbers and the percentage of viable cells from these data.

Another factor to be considered in estimations of this sort is that the relationship between cell viability and productivity may not always be a simple one. For example, several groups have shown that monoclonal antibody production from hybridomas can actually be maximized during the decline phase of the culture due to the release of stored antibody as cells become damaged and permeabilized.

Estimation of the activity of released enzymes such as LDH could be performed on a near on-line basis using automated flow injection analysis and could serve as an effective early warning of cell damage within the fermentor.

Nuclear magnetic resonance technology

Nuclear magnetic resonance (NMR) has recently been applied as a noninvasive technique for examining the status of animal cells growing at high-density in fermentor configurations in which the cells cannot be seen directly. The NMR technique is relatively insensitive and around 10^8 cells/ml are required to generate meaningful signals in a reasonably short time.

Hollow-fibre cartridges have been constructed to operate in the NMR magnetic field and a fermentor containing a high-density culture in macroporous beads has also been employed in this way (Mancuso et al, 1990).

Two main approaches have been employed to estimate cell number and viability. In one approach, intracellular sodium was measured. Since sodium concentration in the medium used was 158 m*M* and the intracellular sodium concentration was only 6.8 mM, the total concentration of sodium in the cellular (extracapillary) compartment of the hollow-fibre fermentor decreased as cell mass increased, thus giving a direct estimate of viable cell number. Intracellular sodium can be specifically estimated since the signal from sodium inside cells shows a considerable shift in the NMR spectrum from that for extracellular sodium (Gupta, 1987).

Another approach is the generation of a phosphorus NMR spectrum. In this case, many molecular species are detectable both inside and outside the cells including inorganic phosphate, phosphate diesters, phosphoenolpyruvate and three different signals from the α, β and γ phosphate groups of nucleoside triphosphates. Intracellular nucleoside triphosphate occurs in the spectrum at levels which are proportional to the viable biomass. However, typical ^{31}P spectra derived from growing cell cultures are complex and may be difficult to interpret. Recently, it has become possible to eliminate the signals from extracellular molecules by applying a treatment known as diffusion weighting to the signals, which permits the elimination of signals from molecules with longer diffusion paths which are characteristic of an extracellular environment (van Zjil et al, 1991).

Good agreement was obtained between the sodium and phosphorus NMR methods of cell number determination (Mancuso et al, 1990).

Laser flow cytometry

In suspension cultures, flow cytometry can provide near on-line monitoring of cells by measuring a number of important parameters including cell viability, cell proliferative capacity and cell product generation capacity.

A variety of fluorescent stains are available for flow cytometric analysis. Of practical relevance for cell culture studies are Rhodamine 123, which is a mitochondrial stain giving a direct indication of cellular metabolic activity, the standard DNA dyes which provide data on the position of the cell within the cell cycle and fluorescent labelled antibody to the required protein product secreted by the cell. This latter directly measures the number of cells which are actively secreting product. Flow cytometry also gives information on cell size within the different classes identified by fluorescence.

The data provided by flow cytometry give new insight into the complexity of the changes that occur in animal cells as they traverse different stages of culture in the fermentor. Cell size, metabolic activity and production capacity can all vary widely at different stages. The associated changes that occur in cellular synthetic pathways can affect not only the yield of product but also its authenticity, underlining again the need for effective monitoring and process control if reliable industrial production is to be achieved. Recent studies suggest that the synthesis of recombinant proteins is linked to the cell cycle and that manipulation of progression through the cycle could represent a viable approach to optimizing expression. Flow cytometry is the most useful tool available for the study of this phenomenon.

Laser flow cytometry has also been used to study the changes in cell viability and metabolic activity produced by hydrodynamic shear forces and by sparging in hybridoma cultures (Al-Rubeai, Chalder and Emery, 1991). It has also been used to monitor the proliferative capacity of Sf9 insect cells in culture, and the progress of infection of these cells by baculovirus (Fertig, Kloppinger and Miltenburger, 1990).

The distribution of cells that are in different phases of the cell cycle represent a very sensitive indicator of the culture conditions since this distribution is one of the first parameters that alters when culture conditions change. In perfused cultures, the cell cycle distribution can be modified by changing the flow of medium in response to flow cytometric data. Flow cytometric analysis could therefore be used to control the culture (Al-Rubeai et al, 1992).

Rapid monitoring of product generation

It would be very useful to be able to continually monitor the yield of protein products from animal cells in culture. On-line methods are not currently available for this, but automated flow injection immunological analysis has been applied to the assay of specific proteins released into the culture medium (Scheper et al, 1990). In this system, culture supernatant is sampled into an external cartridge containing immobilized specific antibody. Concentration of the analyte (antigen) is measured by competition with a calibrated fluorescent-labelled antigen preparation. Results obtained agreed closely with those produced in an off-line ELISA system (Scheper et al, 1990). Although the analysis is discontinuous (analysis time is 15 minutes per sample), it is sufficiently rapid in practice since yield updates at this frequency would easily be adequate to monitor protein productivity in cultured animal cells and to use this information for feedback control and process optimization.

Another immunologically based method, using antibody-coated polystyrene microspheres for rapid antigen capture (the particle concentration fluorescence immunoassay, PCFIA), has also been applied to rapid-process monitoring in animal cell fermentations (Jervis, Lee and Kilburn, 1991). Using this system, assays were complete in 25 minutes.

Rapid HPLC analysis of monoclonal antibody production has also been presented as a useful tool for monitoring the progress and productivity of cell cultures (Holmberg, Ohlson and Lundgren, 1991). With the advent of increasing numbers of rapid, high-performance and highly specific assay methods it is likely that direct monitoring of bioproduct secretion will become widely applied in the near future.

Process parameters for modelling and control

In the absence of fully satisfactory on-line sensors capable of directly monitoring all the relevant parameters in animal cell cultures, several groups have attempted to derive mathematical models which can accurately represent the process from data gathered for those parameters which are readily measurable. In view of the observed great complexity of the nutritional and other requirements of animal cells in culture, a fundamental question concerning this approach is whether the cells' behaviour can ever be adequately described by relatively simple models.

In microbial systems, growth can be described by the Monod model and its derivatives in which concentration of a limiting substrate determines the specific growth rate up to a characteristic maximum value. Recent careful studies using AFP-27-NP hybridoma cells indicate that, at least for this nonproducer hybridoma, a Monod-type model can be adapted to describe cell growth adequately (Frame and Hu, 1991).

Mailly et al (1991) have used ammonium ion concentration as the only measurement to follow the progress of batch and continuous hybridoma cultures. They used a mathematical estimator procedure adapted from techniques previously applied to microbial cultures. In this technique, a Kalman filter approach is employed to resolve differences between observed and measured substrate and metabolite concentrations (Bellgardt et al, 1984).

Using this methodology, it was possible to estimate glucose concentration, cell density and antibody production with reasonable accuracy from measured ammonium ion levels. Further studies are needed to determine what are the best process identifiers for use in this way, how these can best be monitored on-line and how accurate and reliable models and estimators can be derived from this information. Efficient mathematical estimators of this type could be incorporated into fermentor control loops to pilot the process automatically as is already done industrially with some microbial fermentations.

Detailed modelling of cellular metabolism requires a profound knowledge of cellular metabolism and would greatly facilitate process optimization and control and the prediction of the effects of perturbations of the culture conditions. Coupled with adequate computational power, it could also be used diagnostically to identify the causes of and intelligently correct for deviations from a standard culture profile retained in the data base. Detailed modelling and its experimental verification can also provide new insights into cellular physiology.

Several complicated models for animal cells in culture have been published which are claimed to accurately predict various critical aspects of the fermentation such as nutritional requirements and product generation capacity, for example in the case of hybridomas (Barford et al, 1992; Barford and Harbour, 1991; Nielson et al, 1991; Goergen et al, 1991). Most of these are either unstructured or simple structured models since the development of complex structured models for animal cell metabolism remain restricted by limits in current knowledge of cell physiology.

It is important to point out that these models all assume the ideal case in which environment is homogeneous throughout the fermentor (i.e. the fermentor model is unstructured). As has been discussed earlier, this is far from being true for several of the high-density fermentor configurations used for animal cells. In addition to the existence of several intentionally mechanically distinct compartments in some reactor designs, heterogeneity also arises due to the formation of gradients, particularly in plug-flow systems. Also it is difficult to avoid pockets where cells may be anoxic, malnourished or subject to the accumulation of toxic metabolites. This can result in some cells being locally in a declining phase whilst others are still proliferating or are being maintained static but at high viability. Heterogeneity in the fermentor also raises the question of how representative of the whole culture is the data measured and transmitted by a given sensor. These aspects complicate model building, and considerably more complex modelling treatments are therefore required before truly complete models of production in animal cells can be achieved.

4 Adjusting Cellular Metabolism for Optimum Product Yield

One consequence of more intensive monitoring of the different parameters of cell growth and maintenance in fermentors is that it is becoming possible to understand in finer detail the metabolic requirements of animal cells and how these relate to the efficiency of cells as bioreactors for the production of recombinant proteins and other products. With this understanding it is possible to extend attempts to improve the yield and authenticity of the product to include manipulation of the internal mechanisms of the cell in addition to empirical optimization of the cell's external environment through parameters such as fermentor configuration and medium composition. Specifically, information is accumulating on the effect of cellular metabolism on specific product yield, on the generation of undesirable by-products and on the correctness of post-translational modification. An exciting development is the possibility of specifically tailoring aspects of the cell's metabolism to optimize these functions. Useful results have already been achieved even though the detailed metabolism of cells in culture is still far from being understood.

Energy sources and waste products

Mammalian cells in culture use two main energy sources, glucose and glutamine, for the production of ATP and reduced pyridine nucleotides. The proportion of cellular ATP derived from each substrate varies widely with cell type and culture conditions. However, for most types of cell over half of the ATP generated derives from glutamine oxidation. This can rise to very high levels (>95%) under culture conditions in which lactate production is disfavoured (Reitzer, Wice and Kennell, 1979).

Energy derived from metabolism of these compounds is used for the production of new cellular material, for maintenance of cellular integrity and for the synthesis of the required product. In cells engineered for high-level expression of recombinant protein, a major proportion of the available cellular energy source may have to be channelled into the synthesis of product if maximum yield is to be achieved. Intracellular competition for high-energy intermediates and other metabolic precursors may therefore determine product yield (Grummt, Paul and Grummt, 1977), and availability of glucose and glutamine can be a major rate-limiting factor. However, in systems as complex as animal cells, it is not possible to describe the situation so simply because of the pleiotypic effects of apparently simple materials. Thus, in addition to its role as an energy source, glutamine is also essential as an amino donor for several synthetic pathways, including nucleotide synthesis, amino sugar synthesis and the production of asparagine. Glutamine also plays a regulatory role in DNA replication.

ATP and NADPH are produced by the simultaneous catabolism of glucose and glutamine. Glucose may be catabolized by aerobic glycolysis, in which case the end product is lactate, or via the TCA cycle to yield CO_2. The glycolytic pathway yields 2 moles of ATP per mole of glucose consumed whereas the TCA cycle is far more efficient and can yield 36 moles of ATP per mole of glucose. However, only a small proportion (<5%) of glucose is usually metabolized via the TCA cycle in cultured animal cells.

Glutamine catabolism (glutaminolysis) can utilize several different branched pathways in which the glutamine may be converted to lactate, to intermediates such as alanine or aspartate, or completely to CO_2 via the TCA cycle. Yields of ATP per mole of glutamine consumed are between 6 and 9 moles for oxidation to CO_2. Whichever pathway of glutaminolysis is used, between one and two moles of ammonia is released per mole of glutamine consumed (Figure 4.1).

Both lactate and ammonia are toxic to cells, and it is widely supposed that, in addition to nutrient depletion, it is the accumulation of these molecules which cause batch and fed-batch cultures to stop growing and to decline in viability (Dodge, Ji and Hu, 1987).

Numerous studies have been performed to attempt to control the production of lactate and/or of ammonia by limiting glucose and glutamine concentrations in the medium and by adjusting feed rates of these nutrients in fed-batch and perfused systems. A complication of this approach is that a regulatory interdependence appears to exist between the glycolytic and the glutaminolytic pathways (Zielke et al, 1978). The production of lactate can be lowered substantially by using carbohydrate sources other than glucose (Imamura et al, 1982) and also by maintaining a low glucose concentration in the medium of fed cultures (Reitzer, Wice and Kennell, 1979), but both of these approaches to the limitation of lactate production cause increases in glutamine consumption with correspondingly increased levels of ammonia production. Since ammonia is more toxic than lactate, this type of manipulation is not useful. Indeed the opposite approach has been taken, and it has been demonstrated that the maintenance of high glucose concentration by continuous feed drastically lowers glutamine consumption and ammonia release (Glacken, Fleischaker and Sinskey, 1986), although obviously at the price of increased lactate production.

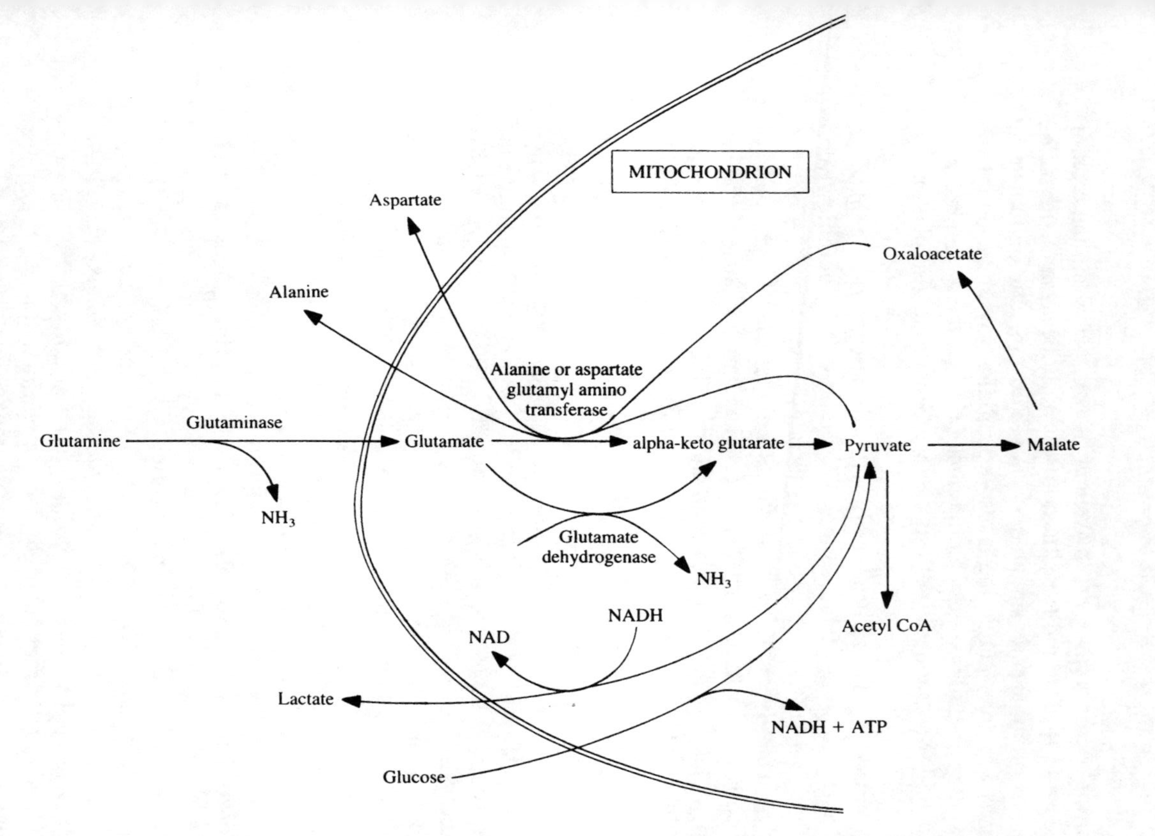

Figure 4.1 Alternative pathways for the conversion of glutamate to alpha-ketoglutarate. Conversion via glutamate dehydrogenase liberates free ammonia whereas the end products when the glutamyl-aminotransferase enzymes are used can be alanine and aspartate.

Recent studies by Butler et al (1991) have shown that ammonium ion concentration in the medium directly affects glucose metabolism. These workers showed that when ammonium chloride (0–6 mM) was added to PQXB1/2 hybridoma cells, glucose consumption and alanine secretion by the cells was doubled. It was hypothesized that ammonia may be detoxified in this system by complexation with α-keto acids such as pyruvate which are produced by glycolysis, and that this may also explain the observed increased release of alanine from cells (Butler et al, 1991). The observation that exogenously added lactate exerted a sparing effect on ammonia toxicity was considered to be compatible with this hypothesis.

Mechanisms of ammonia toxicity

Several possible explanations have been advanced for the toxicity of ammonia in animal cells, including the inhibition of specific enzymes, the alteration of intracellular pH and the disruption of pH gradients, and the imposition of increased maintenance energy demands on the cells.

Ammonia is known to increase intracellular pH (Dean, Jessup and Roberts, 1984; Suput, 1984), which is thought to be due to the fact that un-ionized ammonia can diffuse freely through the cell membrane whereas the charged ammonium ion cannot. Once inside the membrane, the ammonia ionizes by capturing a proton resulting in a local pH rise. Some evidence that un-ionized ammonia is the toxic species, rather than the ammonium ion, is provided by the work of Butler et al (1991), which indicated that the toxicity of ammonium chloride added to the culture medium is enhanced at higher pH values which favour the presence of un-ionized ammonia. However, the possibility that the different medium pH values used (pH 6.8–7.8) might themselves have caused metabolic perturbations and, for example, changed the relative flux in the glycolytic and glutaminolytic pathways should not be discounted.

A second element which is probably implicated in the toxicity of ammonia concerns the active transport of the ammonium ion into the cell by the Na^+/K^+–ATPase pump. The Na^+/K^+–ATPase system uses hydrolysis of ATP to provide energy to pump K^+ ions into the cell to maintain the potassium-rich intracellular environment. It can also be coupled, via the ionic gradient it generates, to systems which control glycolysis, oxidative phosphorylation and intracellular calcium distribution. Accordingly, the pump is intimately associated with intracellular signalling and cell-cycle control (Leiter, Wenner, and Tomei, 1985).

Since hydrated NH_4^+ and K^+ share the same ionic radius, ammonium ions are transported into the cell in competition with K^+. Interestingly, increasing medium potassium concentration was shown to have a sparing effect on ammonia toxicity in this system. The combination of passively entering NH_3 and actively transported NH_4^+ generates intracellular and extracellular pH changes (Martinelle and Haggsrom, 1992). As will be discussed later, pH changes in the Golgi apparatus produced by ammonia can have significant effects on post-translational modification of proteins.

The increase in maintenance energy required to compensate for the wasted pumping of NH_4^+ may also be a major factor in reducing cell viability and product generation capacity. Animal cells typically expend over half of their total energy consumption on maintenance energy, and the Na^+/K^+–ATPase pump is particularly costly in energy and has been estimated to account on its own for the consumption of over 50% of the *total* energy production of rabbit renal cells (Harris et al, 1981). Clearly, signficant waste of metabolic energy in this way could seriously curtail the capacity of the cells to proliferate or to produce protein products.

It has also been suggested that ammonia may be implicated in so-called 'futile metabolic cycles' such as the mitochondrial glutamate dehydrogenase cycle in which ammonium ions are cyclically added to and released from glutamate with a net hydrolysis of ATP, again exerting toxicity by depleting the high-energy phosphate pool (Tagler et al, 1975).

Other studies (McQueen and Bailey, 1990) have shown that addition of ammonium chloride at 10 mM to cultures of TIB 131 hybridomas produced substantial growth inhibition although there was little effect at 3 mM. Addition of ammonium chloride lowered the integral cell yield from both glucose and glutamine, and considerable shifts in the cells' metabolic profile were observed. However, concentrations of ammonium chloride that had inhibited cell growth had no effect on specific antibody production rate or on the characteristics of the antibody produced (McQueen and Bailey, 1990).

Protection against ammonia toxicity

Several approaches have been reported to protect cells from the toxic effects of ammonia.

Removal of ammonia

Perhaps the most obvious approach to lowering ammonia toxicity is to continuously remove ammonia from the medium as it is generated. This is simplified since ammonia can be removed as a gas. Hydrophobic microporous hollow-fibre devices have been used with an acidic stripping solution on the other side of its membrane from the medium to push the equilibrium in favour of diffusion of ammonia out of the medium compartment. Ammonia is then eliminated in the gaseous form from the acid-stripping solution (Hecht, Bischoff and Gerth, 1990; Brose and van Eikersen, 1990). Using this system it was possible to maintain ammonia levels below 0.5 mM.

Another approach has been to remove ammonia as the ammonium ion by diffusion through a cation exchange membrane. In this system, the ammonium ion is deprotonized by an alkaline solution on the stripping side of the membrane, thus promoting increased flux of ammonium ions out of the medium. Gaseous ammonia can subsequently be irreversibly removed from the stripping solution by sparging with inert gas (Thommes et al, 1992). A possible limitation of this method might be the loss of other cations through the relatively unselective cation-exchange membranes currently available.

Another method of ammonia removal which has been shown to be effective is absorption using cation exchange resins (Iio and Moriyama-Takashima, 1985). In these studies, dialysis tubes packed with the sodium form of clinoptilolite, a natural cation exchanger, were immersed in the medium and considerable reductions in ammonia concentration were obtained.

Reducing ammonia production

As we have seen, reducing ammonia production is not simple because of the interactive nature of the different ATP-generating metabolic pathways in cultured animal cells. One strategy to minimize the quantity of ammonia released per mole of ATP produced, discussed by Glacken (1988), would be to manipulate mitochondrial glutamate metabolism to favour the conversion to α-ketoglutarate via glutamate aminotransferase rather than via glutamate dehydrogenase (see Figure 4.1), since in this case the end products of the reaction are the relatively nontoxic alanine and aspartate rather than free ammonia. Although this reaction sequence produces less ATP per mole of glutamine metabolized (9 moles of ATP per mole of glutamine compared with 9, 12 or 15 moles of ATP produced by the dehydrogenase route, depending on whether the end products are lactate, pyruvate or acetyl-CoA), the ratio of ammonia produced to ATP generated is lower (Glacken, 1988 and see Table 4.1). The inhibition of glutamate dehydrogenase has been proposed by Glacken (1988) as a means of achieving increased flux through the glutamate aminotransferase pathways. Fumarate and pyridoxal phosphate were considered possible inhibitors for use in culture medium. Glacken also pointed out that leucine activates glutamate dehydrogenase and that the concentration of this essential amino acid should therefore be kept as low as possible in the culture medium consistent with acceptable cell growth and product yield.

It should also be possible to reduce ammonia production by lowering the concentration of glutamine in the medium, and several studies have

Table 4.1 *Alternative pathways for glutamine utilization: relative yields of ATP and relative ammonia production*

Metabolic Pathway	End Product of Glutamine Metabolism	Moles ATP produced per mole glutamine	Moles ammonia produced per mole glutamine	Ratio of Ammonia:ATP
Amino transferase route	Alanine	9	1	0.11
	Aspartate	9	1	0.11
Glutamate dehydrogenase route	Acetyl CoA	15	2	0.13
	Pyruvate	12	2	0.17
	Lactate	9	2	0.22

shown that this is indeed the case. The use of glutamine-limited fed-batch cultures can reduce both ammonia production and overflow production of the nonessential amino-acids, alanine, proline, ornithine and asparagine. More efficient utilization of both glucose and glutamine as an energy source can be obtained (Ljunggren and Haggstrom, 1990). In these studies, it was postulated that reduced ammonia production was due to reduced availability of substrate for glutaminase and that reduction in overflow production of nonessential amino acids was due to reduced availability of glutamate for transamination reactions. However, as glutamine concentration is lowered to around 1 mM, significant reduction in growth rates occurs.

As already mentioned, increased glucose availability has a sparing effect on glutamine consumption, and glucose utilization decreases as glutamine concentration decreases (Zielke et al, 1978). In principle, if glycolysis could be increased sufficiently to satisfy all of the cells' energetic needs, the glutaminolytic pathway could be shut down completely as long as essential intermediates normally supplied by this pathway, such as aspartate and asparagine, are included in the medium formulation (Glacken, 1988).

However, as discussed, accentuating glucose metabolism to limit glutaminolysis would produce significant inhibition of both growth and product yield since very high levels of lactate would accumulate. This apparent impasse is due to the inefficiency of glucose oxidation by animal cells in culture. For example, in typical cells, 80% of glucose metabolized is incompletely oxidized and accumulates as lactate and only about 5% enters the TCA cycle and is oxidized to CO_2 (Zielke et al, 1978; Reitzer, Wice and Kennell, 1979). Strategies which could increase the percentage of glucose flowing into the TCA cycle would be expected to enable the

Table 4.2 *Possible reasons why the proportion of glucose oxidized via the TCA cycle in animal cells is low*

Possible Cause	Possible Strategy for Modification	Reference
Deficiency in transfer of reducing equivalents from cytosol to mitochondria possibly due to deficiency of shuttle intermediates	Provision of adequate levels of shuttle intermediates such as aspartate, malate or citrate in the culture medium	Nakano, Ciampi and Young (1982) Glacken (1988)
Damage to Krebs cycle enzymes due to high levels of superoxides in relation to available levels of superoxide dismutase. Energy is therefore diverted into repair mechanisms	Maintenance of low redox potential and provision of precursors for damage repair	Glacken (1988)
Reduced pyruvate dehydrogenase (PDH) activity	Activation of the PDH complex by dichloroacetate treatment	Murray, Dickson and Gull (1992)

cells' energy requirements to be met in the absence of glutamine and without massive production of lactate.

It is not clear why the TCA cycle is so little used by cultured cells, although several hypotheses have been advanced as indicated in Table 4.2. Several theoretical solutions can be conceived on the basis of these hypotheses and these are also listed in Table 4.2.

Are there metabolic limitations beyond lactate and ammonia toxicity?

Although it is well established that both lactate and ammonia inhibit productivity in many cell lines (Glacken, Fleischaker and Sinskey, 1986; Reuveny et al, 1986) it is not formally established that these compounds are the factors that cause growth arrest and cell death in typical cell cultures. Schumpp and Schlaeger (1992) recently studied the growth of double mutants of HL-60 human promyelocytic leukaemia cells resistant to lactate (60 mM) and ammonia (4 mM) and compared them with the growth of wild-type H2-60 cells in medium containing added lactate and ammonia.

As expected, the mutants grew equally well in the presence or absence of lactate and ammonia concentrations that were nonpermissive for wild-type cells. However, the expected increase in maximum cell number of the mutants in standard medium was not observed. Since these cells were unlikely to have been inhibited by accumulated ammonia or lactate, the implication is that some other metabolite(s) inhibit cell growth in these conditions. Although growth of the mutants was similar in medium with or without added lactate and ammonia, the metabolic profile obtained in control medium were quite different from those obtained in the presence of lactate and ammonia. One characteristic of the resistant clones was increased alanine secretion, suggesting perhaps that the glutamate aminotransferase pathway for glutamine metabolism was favoured by these cells.

Growth in glutamine-free medium

Another contribution to the production of ammonia in culture is the spontaneous decomposition of glutamine to ammonia and pyrrolidone carboxylic acid. Lower glutamine concentrations would reduce ammonia production by this route, but glutamine remains an essential medium additive for several types of cell. Bell et al (1991) showed that hybridoma cells could be rendered independent of exogenous glutamine by transfection with the glutamine synthetase gene. The engineered hybridoma did not accumulate ammonia in the medium when grown in glutamine-free medium whereas the wild-type hybridoma grown in 2 mM glutamine did accumulate ammonia. Interestingly, the engineered cells produced less lactate per mole of glucose metabolized, but the mechanistic basis for this has not yet been determined.

Endogenous glutamine synthetase has been shown to be induced in several cell types by glutamine deprivation, and this may be the main mechanism by which cells can adapt to glutamine-free medium. Thomas, Jenkins and Butler (1991) demonstrated that this induction occurs in both McCoy and MDCK cells. However, though McCoy cells could be adapted to grow well in the absence of added glutamine, MDCK cells were still incapable of growth in glutamine-free medium despite the efficient induction of glutamine synthetase. Further studies indicated that this was due to poor uptake of glutamate, the glutamine synthetase substrate, by MDCK cells, and that this deficiency rather than the lack of the enzyme was the primary barrier to growth in glutamine-free medium.

Effect of culture conditions on protein glycosylation

Glycosylation represents the most frequent post-translational modification of proteins and has profound effects on their functionality. The carbohydrate moiety added may constitute a large proportion of the finished protein, about 10% by weight in the case of IgM and 40% in the case of erythropoietin (Goto et al, 1988). The presence of this large carbohydrate structure on the surface of a protein is frequently determinant for critical properties such as solubility, stability, biological recognition, biological clearance, and so on. Correct glycosylation is fundamental to the production of structurally and biologically authentic recombinant glycoproteins.

However, in contrast to the translation of the gene sequence into

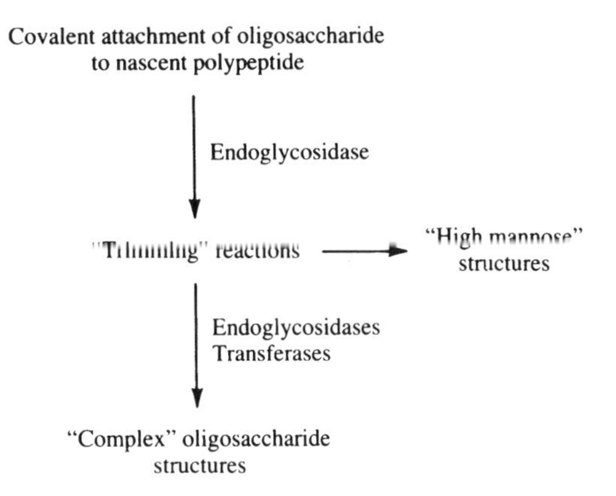

Figure 4.2 Summary of protein *N*-glycosylation pathways: The stepwise glycosylation occurs in several subcellular compartments and involves multiple enzymes. Initial attachment of the oligosaccharide to the nascent peptide occurs in the ER and involves oligosaccharide transferase, α-glucosidase I, α-glucosidase II and ER α-1,2 mannosidase. Processing continues during transit of the Golgi apparatus with Golgi α-mannosidase I (*cis*-Golgi), *N*-acetylglucosaminyl transferase I, Golgi α-mannosidase II, *N*-acetylglucosaminyl-transferase II, fucosyl transferase (*medial*-Golgi), galactosyl transferase (*trans*-Golgi), and sialyl transferase (*distal*-Golgi) followed by exit from the Golgi network.

amino acids, whose precision is assured by reading the genetic template during protein synthesis, glycosylation is entirely dependent on the co-ordinated, sequential action of various enzymes in different cellular compartments (Figure 4.2). Without the constraints provided by a template to determine structure it is not surprising that the glycosylation patterns of protein can be significantly influenced by alterations in the cellular environment.

The consequences of incorrect glycosylation can either be the transport and secretion of aberrant glycoforms which would compromise the safety and efficiency of the product, or, if the aberrantly glycosylated protein folds incorrectly or tends to aggregate, the protein can become bound to chaperone proteins such as BIP and sequestered in the ER (Kassenbruck et al, 1988).

Also important is the observation that the same protein expressed in different cell lines can be glycosylated in significantly different ways. It has been known for some time that the glycosylation of viral protein is highly host-cell dependent (Hsieh, Rosner and Rabbin, 1983). However, Kagawa et al, 1988 have showed that this also applies to recombinant proteins. They showed that human beta interferon expressed in CHO cells carried the same oligosaccharide structures as natural interferon beta. This consists predominantly (> 80%) of bi-antennary complex-type sugar chains, the remainder being 2,4 and 2,6 branched tri-antennary structures. However, human beta interferon derived from C127 mouse epithelial cells and PC8 human adenocarcinoma cells differed considerably in oligosaccharide structure. In particular, both secreted interferon containing the Gal α 1 —> 3 Gal sequence against which naturally occurring antibody circulates in all human sera (Kagawa et al, 1988). Obviously synthesis of this carbohydrate epitope totally compromises the safety and efficacy of beta interferon produced in these cells.

In either case, efficiency of product generation can be dramatically curtailed. The effects of cell culture environment on protein glycosylation have recently been reviewed by Goochee and Monica (1990).

Factors influencing glycosylation in cultured cells

Glucose availability

Several studies have shown that glucose limitation results in incomplete protein glycosylation. This manifests in two ways (Elbein, 1987). In some cases, abnormally small dolichyl precursor oligosaccharides are added to asparaginyl glycosylation sites in the initial cotranslational glycosylation step [for example $(\text{Glucose})_3$, $(\text{Mannose})_5$, $(N\text{-acetylglucosamine})_2$ instead of the normal $(\text{Glucose})_3$, $(\text{Mannose})_9$, $(N\text{-acetylglucosamine})_2$ moiety]. In other cases, some normally glycosylated *N*-linked glycosylation sequences remained completely unglycosylated in glucose-starved cells (Stark and Heath, 1979; Davidson and Hunt, 1985). These effects observed in glucose starvation conditions are probably related to both overall energy depletion of the cells and to a specific lack of appropriate oligosaccharide precursors. Although glucose starvation would not generally be expected to be a problem in most animal cell fermentations, local glucose depletion could occur in imperfectly mixed, high-density cultures and the protein products of such cultures could therefore be heterogenous in terms of glycosylation.

Ammonia and other amines

Ammonium chloride added to cell cultures results in glycoproteins deficient in terminal sialation, a reaction which is catalyzed by sialyl transferase in the distal Golgi compartment (Thorens and Vassalli, 1986; Oda et al, 1988). As already discussed, the toxicity of ammonia is largely due to its accumulation (by passive diffusion of NH_3 and by active transport of NH_4^+) in pH sensitive intracellular compartments (Dean et al, 1984; Suput, 1984). Disruption of the pH in the normally acidic distal Golgi is thought to be responsible for inhibition of the sialyl transferase. Other amines such as Tris may have similar effects.

This observation underlines the importance of the control of ammonia accumulation in cultures since, in addition to affecting cell yield and product generation, ammonia toxicity can also affect product quality and authenticity.

Effect of hormones, vitamins and growth factors

Multiple examples exist in which treatment of cells in vitro with various hormones, vitamins, differentiation factors, and so on, results in altered glycosylation patterns in glycoproteins secreted by the cells (reviewed by Goochee and Monica, 1990). It seems most likely that these changes generally result from the induction or repression of the enzymes involved in protein glycosylation. For example, Wang, O'Hanlon and Lau (1989) have shown that dexamethasone treatment of hepatocytes and hepatoma cells results in a three- to four-fold increase in sialyltransferase activity and that this was due to increased transcription of the sialyl transferase gene.

Changes in glycosylation patterns

There are many examples in which glycosylation patterns have been shown to change during the course of an animal cell fermentation or when culture conditions have been altered, for example in moving to the next stage in scale-up. This phenomenon has recently been reported in the production of interferon gamma in CHO cells (Curling et al, 1990) in which heterogeneity of glycosylation of the product occurred as a result of the proportion of interferon molecules which remained unglycosylated increasing during the course of the culture. Cadeiro and Curling (1992) have also reported variations in the glycosylation of an IgM monoclonal antibody, depending on the availability of glucose in the medium.

These observations underline the need to verify glycoform structure when medium components are changed and particularly if production is changed from serum-containing to serum-free medium. Batch to batch variation in serum hormone or vitamin levels might also effect the quality of product through this mechanism. Such factors would be difficult to control in serum-containing medium and difficult to identify as critical components in serum-free medium.

Effects of glucose on other post-translational modifications

Sugiura (1992) has recently published data on the production of protein C in CHO cells. Protein C is a plasma protein with anti-coagulant activity which requires vitamin-K dependent gamma-carboxylation at specific *N*-terminal glutamic acid residues for its biological activity. In this study, glucose levels of 3 mg/ml and above were shown to impair gamma carboxylation of protein C and to inhibit protein C production. This phenomenon is unexplained at present.

The strategy proposed by Sugiura to obtain maximum yield of protein C involved manipulating cellular metabolism by separating the process into two phases, the growth phase and the production phase. In the growth phase high levels of glucose levels were used whereas for production, glucose levels were maintained below the critical value of 3 mg/ml to obtain the highest expression rate and optimal post-translational processing.

Other culture parameters that affect yield

Optimizing amino acid composition of the medium

Many investigations have shown that specific amino acids may become depleted in batch cultures, and replenishment of these can result in enhanced productivity. Interestingly, cellular amino acid metabolism does not remain constant during the life of a culture, and several metabolic phases may be traversed. Product generation is apparently uncoupled from growth for many cell types, and this may be reflected in different amino-acid requirements at various points in the process.

Recent detailed studies on the consumption of individual amino acids during hybridoma growth and antibody production revealed a complicated phasic pattern of amino acid utilization. Some amino acids were important for growth but not during the antibody-production phase whereas others could be omitted during growth but profoundly affected antibody yield (Bell et al, 1991). It appears probable therefore that there will be competition for amino acids used for cell growth and for product synthesis and that the amino acid requirements for these two phases of culture may be different. How cells control switching between proliferation and product synthesis is not known, but it is probable that fine tuning of the medium amino acid composition is one factor that could be used to facilitate or favour one of these alternatives.

As mentioned earlier in this chapter, the concentration of leucine in the medium may be of particular importance in this connection since glutamine acts as an activator of glutamine dehydrogenase. Leucine could thus favour the catabolism of glutamine via the dehydrogenase route with the release of an additional molecule of ammonia which would accentuate the accumulation of ammonia in the culture.

Competition for energy pools

Several studies indicate that cell growth and product generation may also be mutally competitive at the level of utilization of high-energy intermediates. For example, stimulation of lymphocytes with concanavalin A results in a 35% drop in the intracellular ATP pool, suggesting that DNA synthesis may represent a considerable drain on the energy supply in the cells (White et al, 1989). This type of observation led Modha, Whiteside and Spier (1992) to try to manipulate cellular metabolism to switch in favour of product generation by inhibiting DNA synthesis to make more of the cell's energy pool available for the synthesis of product. A variety of DNA synthesis inhibitors were shown to produce an increase in intracellular ATP concentration and in cell-specific antibody production rates, although it was considered unlikely that there was an increase in antibody productivity per unit biomass.

Although these areas are the subject of intensive study, much remains to be done before the factors controlling the switch(es) from cell growth to product synthesis are understood (Modha et al, 1992).

Protease release by cells

Many types of animal cells secrete protease into the culture medium. In the absence of the protease inhibitors in serum or of other exogenously added protease inhibitor, the degradation of protein products by these enzymes can be a serious problem. The types of proteases produced have not been fully characterized, although plasminogen activators of the uPA type occur frequently in mammalian cells and are generally secreted at higher levels by transformed cells than by normal cells (Hatcher et al, 1977). Insect cells can also secrete troublesome levels of proteolytic activity, although the nature of the enzymes involved in this case has been very little studied.

Protease secretion by cells depends on the culture conditions. It has been known for some time that low protein medium and high culture pH are both factors which favour protease production (Schlaeger, Eggimann and Gast, 1987).

More recent studies have shown that the induction of protease may be a direct result of amino acid starvation. Froud et al (1991) described the production of the HIV envelope glycoprotein, gp120, in CHO cells amplified for gp120 production using the glutamine synthetase system. In these studies, gp120 was consistently produced as a truncated form due to a specific proteolytic cleavage occuring between residues 285 and 286 of the mature protein. Analysis of the kinetics of production of the nicked form showed that the appearance of the protease coincided with the depletion of several amino acids in the culture medium. Use of a fed-batch system to maintain levels of the critical amino acids (asparagine, glutamate, aspartate and serine) prevented the appearance of proteolytic activity and permitted isolation of completely uncleaved gp120. This apparent stimulation of protease activity (it is not known whether the increase is due to induction or increased secretion of the protease, or to release following cell lysis) may be a general phenomenon associated with nutrient depletion and should be investigated in other systems in which proteolytic cleavage is a problem.

Another example of proteolytic activity stimulated by shifts in cellular metabolic patterns caused by changes in culturc conditions has been reported for recombinant BHK 21 cells secreting IL2 (Sugimato, Lind and Wagner, 1992). In this study, the transition from microcarrier to suspension culture resulted in the intense stimulation of protease secretion to a point where over a third of the product was lost by proteolytic degradation. These studies probably reflect the reported involvement of proteases in the attachment and mobility of anchorage dependent cells on surfaces (Goldberg and Dice, 1974).

Overall it appears clear that the production of proteases is another of the animal cell's repertoire of metabolic responses to its environment and as such is amenable to modulation by manipulation of the culture conditions employed. As illustrated in the examples given, control of proteolysis may be a critical element in developing a viable production process.

Osmotic shock

Several apparently unrelated factors all appear to stimulate monoclonal antibody production from hybridomas by a mechanism which remains poorly understood. A unifying element appears to be that the treatments involve subjecting the cells to stresses in the culture conditions by applying hyperosmotic shock, pH change, temperature change, and so on. (Oyaas et al, 1989). When cells are exposed to such stresses, an inverse relationship between growth rate and antibody yield is observed. Although specific antibody production rate increases, cell density generally decreases so that overall antibody yield is usually not greatly improved.

It has been suggested that stressing the cells may influence the cascade of events that results in a switch of cell metabolism from a cell proliferation mode to a product generation mode. In the case of hybridomas, an important aspect may be the stimulation of release of preformed antibody from the cells (Mohda et al, 1992).

Butyrate addition

Addition of sodium butyrate in the 0.5–2.0 mM concentration range has been shown to have a wide range of effects on cells in culture (reviewed by Kruh, 1982). These effects include enhancement of the yield of several secreted protein products (Dorner, Wasley and Kaufman, 1989). Examples of this include alpha interferon from Namalwa cells (Johnston, 1980), monoclonal antibodies from hybridomas (Birch and Boraston, 1987) and factor VIII from recombinant CHO cells (Ganne et al, 1991).

Studies on gene expression in Factor VIII-producing CHO cells indicated that butyrate treatment induced several classes of mRNA, including that for Factor VIII. It has been suggested that butyrate may also induce the expression of chaperone proteins such as grp 78 and grp 94 which are thought to play a role in the transit of secreted proteins through the ER (Dorner et al, 1989). The effect of chaperone proteins on the yield of secreted proteins is discussed briefly in Chapter 2.

Sodium butyrate addition is a simple and inexpensive process which has given useful yield improvements (typically 3–5 fold) with several high-value protein products on the industrial scale.

General remarks on productivity

The productivity of animal cell lines as bioreactors has been hugely improved by obtaining high-level transcription, and the factors governing efficient processing and secretion of proteins are now under intense investigation, promising to lead to further productivity gains. So far, studies on the manipulation of overall cellular metabolism to increase yield have been largely empirical, but substantial improvements in product output and authenticity have nevertheless been achieved. It seems clear that the integration of improved basic knowledge of cellular metabolism with the wealth of accumulated empirical information available will lead to further major gains in yield.

5 Downstream Processing

All of the procedures and methods discussed so far concern how animal cells can be grown in quantity and engineered to synthesize the maximum possible amounts of protein product required from them. In this section we turn to the methodologies applied to the recovery of the protein product from the cell cultures and to turning it into an acceptable pharmaceutical dosage form. The whole of this operation is known as downstream processing (DSP).

The inclusion of a section on downstream processing in a discussion of animal cells as bioreactors is fully justified for several reasons. First, it is usually the effectiveness of the downstream processing which determines the economic viability of the process since it accounts for at least 70–80% of the overall costs. Second, downstream processing plays a determinant role in assuring the safety of products derived from animal cells. Finally, the success of the overall production process depends to a large extent on the successful integration of the expression system used, the cell culture system used and the processing methods applied to produce the finished product.

Although they may not always be easy to define, interactions exist between all of these stages, and early evaluation of these greatly facilitates successful process design. Too frequently in the past biotechnology companies have developed new cell systems for producing original products and have then adapted standard purification schemes to suit. However, there are potentially great savings in time and cost if process design is performed in an integrated manner, for instance, taking into account before the cell culture system is finalized the impact of medium composition or possible cell lysis on the quality of the product stream going for purification. The availability of computer simulation techniques for process design greatly facilitates this type of study.

Downstream processing objectives: production of safe and effective biopharmaceuticals

Biopharmaceuticals are produced from animal cells because of the fidelity with which animal cells can produce authentically processed human proteins. However, the biological and biochemical similarity to human cells which makes this possible also means that animal cells may contain factors which could be hazardous if they persisted in the final dose form. These include animal proteins derived from the production cell or from medium or process additives, viruses, and DNA of cellular oncogene or viral origin. These potential biological hazards are discussed in Chapter 6. In addition, chemical contaminants such as antifoams, antibiotics, leached chromatographic ligands and so on could also persist in the final product. The objectives of downstream processing is to eliminate all of these hazards in a reproducible way to arrive at a final dosage form of acceptable purity which is fully characterized in terms of the authenticity and potency of the required product and the identity and concentration of any possible contaminants. Downstream processing is thus the primary guardian of the safety, purity, potency and consistency of the product as required by regulatory agencies. Because of their high costs, the downstream processing procedures will also most likely determine the economics of the overall process.

Challenges in the purification of products from animal cells

The nature of animal cells and the techniques required for their growth has major impact on the downstream processing procedures required. A primary factor is their need for complex medium which may contain high concentrations of added protein as discussed in Chapter 3. This can provide a high level of background protein, often globally similar in physicochemical properties to the product which must be separated. As mentioned earlier, this situation can be radically improved by the design of an effective serum-free or protein-free medium adapted to the nutritional needs of the production cells.

Since the level at which proteins are secreted by animal cells remains low in most cases, the harvested supernatant contains only a low concentration of the secreted product (generally levels are well below 100 mg/l in stirred-suspension cultures, rising to 5–700 mg/l for the best performing hybridomas). Purification techniques usually work less efficiently in this situation and, indeed, if the required product is present at too low concentration or as a very low percentage of the total protein, purification may not be practical either technically or economically.

The situation can be improved if cells are grown at higher density so that higher product concentration can be achieved. However, as already discussed, high-density cell reactors are difficult to operate with culture conditions that are homogeneous and optimal throughout the fermentor, and leakage or lysis of damaged cells may complicate purification by adding more cellular protein to the purification feed stock. Some of the released cellular proteins may have proteolytic activity and cause degradation of the product.

A particular difficulty can be the generation in the cell culture of closely related variant forms of the required protein. This can be due to differential glycosylation, as discussed in Chapter 4, in which changing culture conditions during the fermentation run results in the synthesis of a preponderance of different glycoforms at different stages. It may also result from partial proteolytic cleavage due either to the variable activity of cellular processing enzymes, the induction of extracellular protease or to cell lysis. Variability in other types of post-translational modification has also been reported (Mikkelsen, Thomsen and Ezban, 1991).

In still other cases, interaction can occur between the product and macromolecular medium components. This has proved very troublesome in the case of tPA production in serum-containing medium. High molecular weight complexes (95–125 kDa) are produced between tPA and several serum proteins. These complexes are inactive but are still recognized by anti-tPA antibodies and by lysine affinity columns and so co-purify with tPA in these widely used purification systems (Lubiniecki et al, 1989; Cartwright, 1992).

In all of these situations, purification can be complicated extremely by these 'difficult' impurities which have some of the properties of the authentic product and which therefore behave similarly in their binding to chromatographic supports, but which are unacceptable in the final product.

Because the required protein itself is usually of a labile nature, only mild purification techniques can usually be applied. This is only poorly compatible with the general requirement to produce a sterile apyrogenic product and also to favourize the inactivation of any infectious agents that could be present. Some characteristic problems which need to be addressed in the purification of protein from animal cells are summarized in Table 5.1.

Design of the extraction and purification process

The downstream processing of biologicals is typically broken down into a number of phases as summarized in Figure 5.1. Detailed discussion of the individual unit operations is beyond the scope of this work, and the reader is referred to the extensive review by Rosevear and Lamb (1988). However, the following points are worthy of particular note:

Primary separation

The separation of supernatant from animal cell culture obviously poses various problems, depending on the type of fermentor in use. The systems in which cells are retained or entrapped in the fermentor or are attached to surfaces produce product streams which do not contain high levels of particulate material, although some detached cells and debris from lysed cells are still present. High-density cell cultures may contain

Table 5.1 *Particular problems which may be associated with the downstream processing of products from animal cells and possible solutions. Note the interactive nature of the different phases of the production process which are nevertheless frequently designed and optimized separately.*

Potential Technical Problems	Possible Solutions
• Production cells fragile and easily lysed	Rapid separation of product from cells Avoid conditions which favour cell lysis either in culture or during extraction
• Low product purity	Product purity is vastly improved by the use of serum-free or protein-free medium. Requirement for exogenous protein is usually reduced in higher density cultures.
• Low product concentration	Usually only mg/litre - but higher yielding cells or higher cell density fermentors may improve this.
• Possible production of minor variants of the required protein	Use cells with post-translational processing capacity able to match transcription and translation rates. Maintain constancy of culture conditions avoiding stress and nutrient limitations. Rapid purification from the culture supernatant.
• Lability of protein product	Can be improved by reducing the residence time in the fermentor, e.g. in perfusion systems with continuous product harvest. Possible use of stabilizers, such as protease inhibitors, during the process Limits the choice of purification procedures and other treatments that can be applied to the product.
• Potential contamination by hazardous biological agents	Development of a validated process which assures clearance of such agents. Choice of clearance procedures is aided by proper characterization of the production cells.

proportionately more cell debris than lower density, less stressed cultures.

A potential problem in the separation of animal cells from supernatants is their relative fragility. Cell disruption during the primary separation process must be avoided since it results in added protein load for the purification process to cope with and there is also the possibility that released proteases could damage the product and so reduce yield and further complicate purification.

In practice, continuous-flow centrifugation is widely used for primary

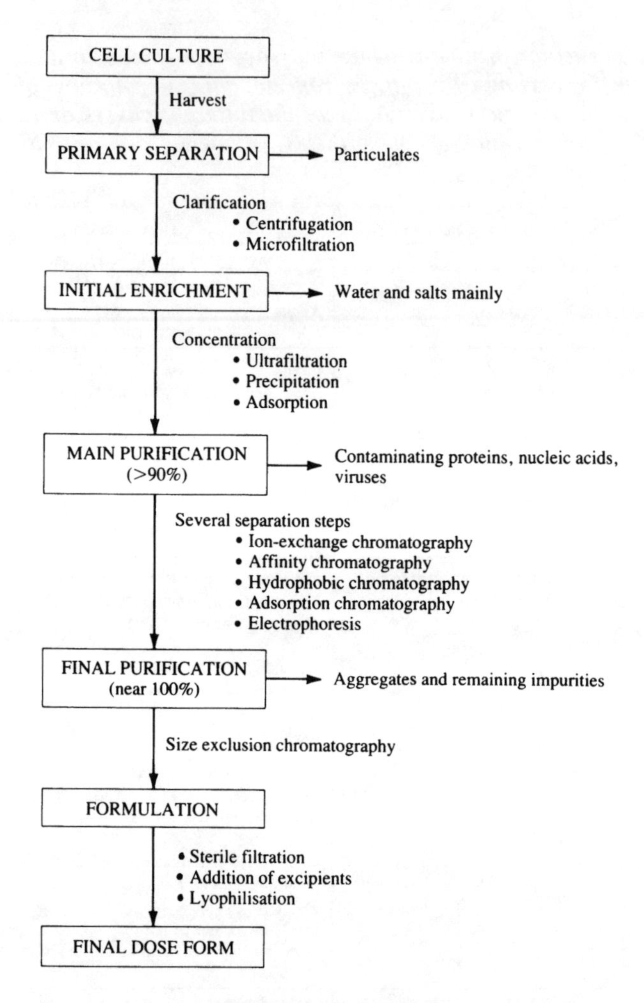

Figure 5.1 Generalized flow sheet for downstream processing steps applied to recombinant proteins from animal cells

separation and provides satisfactory results. A limitation of centrifugation is its relative inefficiency in the removal of fine debris. An alternative procedure, which is more effective for fine debris removal, is crossflow microfiltration. Both systems can be operated at high flux under contained conditions.

Initial enrichment

The major contaminant in culture supernatants is water, and it is critical that water removal be accomplished early in the processing to facilitate

Table 5.2 *Advantages and disadvantages of the different unit operations applicable to the initial enrichment phase*

Operation	Advantage	Disadvantage
Salt Precipitation	Widely applicable Gives some purification High concentration factor	Poor yield Need to remove precipitant Difficult on large scale Inefficient in dilute protein solutions
Solvent Precipitation	Widely applicable Gives some purification Inactivates some viruses High concentration factor	Poor yield Problems of solvent disposal/ recovery Inefficient in dilute protein solutions Denaturation of some proteins
Aqueous liquid/ liquid extration	May give considerable purification Tolerates particulates Mild	Polymers used are expensive Poor concentration factor
Adsorption	Simple Can give substantial purification Can give high concentration factor	Empirical Recovery from adsorbent can be difficult Input material must be adjusted to appropriate adsorption conditions
Ultrafiltration	Very mild treatment Nothing is added to the product stream Easily containable Easy to scale-up Excellent concentration factor Can be used to change ionic conditions (diafiltration)	Minimal purification obtained Subject to fouling

fluid handling, to reduce scale of subsequent unit operations and to enhance the stability of the product.

The main techniques available to achieve this are precipitation of product by solvent or salt, adsorption, liquid–liquid extraction or ultrafiltration (Cartwright and Duchesne, 1985; van der Morel et al, 1985; Rosevear and Lamb, 1988). Ultrafiltration is now almost universally employed for this operation because of its rapid throughput, capacity to operate under contained conditions and its versatility (Table 5.2).

Ultrafiltration membranes have a large porosity bandwidth, therefore, to avoid losses it is usual to concentrate protein using membranes with a nominal cut-off of 10 kDa. Problems of membrane fouling by particulates can be largely overcome by the use of tangential flow ultrafiltration configurations such as the Amicon hollow-fibre system.

In addition to cell debris, membrane fouling can also be caused by process additives such as antifoam or hydrophobic polymers. This problem is particularly pronounced with hydrophobic membranes. Soluble proteins in the product stream can also cause difficulties, especially if they tend to aggregate as they become more concentrated. Albumin in particular has been shown to reduce ultrafiltration performance by binding to the membrane (Matthiasson, 1983).

Main purification

The main purification involves a series of selected, complementary purification steps of the highest resolving power possible with the aim of producing the required protein in a highly purified state (98–100% purity). Effectively, this means that the protein is pure by most of the physicochemical tests that can be applied. The aim is to arrive at this state of purity in the lowest possible number of steps since every step inevitably involves some loss of product yield, and overall yield declines exponentially as the number of steps increases. Appropriate design of the purification sequence is thus critical and represents the most unforgiving part of the overall production process.

In practice, the techniques used in this phase are essentially chromatographic, and a very wide range of adsorbants are currently available which separate according to different physicochemical properties. Because of the high value of biopharmaceuticals derived from animal cells, expensive high resolution chromatographic medium including affinity supports and HPLC systems are frequently applied to these purifications (Cartwright, 1987). Table 5.3 provides an overview of the characteristics of some of the currently available chromatographic techniques.

Final purification

Final purification or polishing steps are required to remove additives which may have been used during the process (such as protease inhibitors to protect the product during purification), to remove ligands which may have leached from chromatographic supports during the process, to eliminate any aggregates of the required product which may have formed and sometimes to change the buffer to that required in the final dosage form. Gel filtration is frequently employed at this stage.

Table 5.3 *Selected chromatographic methods for application to the purification of proteins*

Chromatographic Method	Criteria of Separation	Advantages	Limitations
Size exclusive chromatography (gel filtration)	Molecular size and shape	Mild Selective Can be used to change solvent Useful for 'polishing'	Dilutes
Ion-exchange chromatography	Net charge	Highly selective High capacity Can concentrate product	Feed must be at correct pH and I
Non-defined adsorption chromatography (e.g. various proteins on controlled pore glass, hydroxyapatite, etc)	Unclear and complex	Can be highly selective High capacity	May be difficult to apply reproducibly
Hydrophobic interaction chromatography	Presence of non-polar 'patches'	Most effective at high ionic strength Complements ion-exchange	Dilutes
Metal chelate chromatography	Presence of accessible chelating groups - notably histidine	Highly selective Can be utilized in affinity tagging techniques	
Dye ligand chromatography (pseudoaffinity chromatography)	Complex (ionic and hydrophobic)	Highly selective	Ligand leaching (chemical toxicity)
Bioaffinity systems: Lectins Protein A Protein G Enzyme cofactors, etc	Specific protein: protein interactions	Extremely specific	Ligand leaching (biohazard) Cannot be sterilized High affinity may make elution without loss difficult
Immunoaffinity systems	Antibody: antigen binding	Extremely specific	As above (biohazard)

Current concerns in the downstream processing of protein biopharmaceuticals

The primary aim of the downstream processing operation is to provide a product which is pure and safe. As will be discussed later in this chapter, very high protein purity is relatively easily obtained with modern separation techniques. Several concerns peripheral to the question of purity still persist, however, including

- maintaining the whole process under hygienic, near sterile operating conditions and free from pyrogen build-up.
- the problem of ligands leaching from the chromatographic supports into the product with toxicological consequences.
- the effectiveness of the purification scheme in eliminating contaminating viruses and residual DNA.
- the possible release of biological agents when animal-derived protein affinity ligands such as monoclonal antibodies are used.

These points will be briefly discussed in turn.

'Hygienic' operation

Biopharmaceuticals must be processed under conditions which do not allow the growth of micro-organisms or the accumulation of pyrogens and which give a sterile final product.

It is not practical to conduct the whole operation under strictly sterile conditions. However, all reagents and equipment used must be sanitized and depyrogenated using methods approved by the regulatory authorities, and the plant and operating procedures employed must be of such quality as to prevent the ingress of adventitious agents. In practice, this requires a rigorous cleaning-in-place (CIP) regime to be applied to the purification plant between runs, and this implies effective regeneration of all chromatographic supports to remove bound contamination, removal of pyrogens throughout the system and complete sterilization. Steam sterilization is the preferred method and is applied whenever possible. Although steam-sterilizable ultrafiltration membranes are available, many of the chromatographic supports used will not tolerate steam treatment and chemical sterilants are widely used. Depyrogenation involves washing with cleaning solutions such as sodium hydroxide (>0.1 M) or acid ethanol, both of which have been validated for their capacity to eliminate pyrogens. Again, these treatments can be deterious to chromatographic media. In the case of protein-affinity ligands, such vigorous treatment is obviously out of the question. In such cases, sanitization by treatment with milder depyrogenation agents followed by copious flushing with sterile pyrogen-free buffer are applied (Boschetti, 1991).

Ligand leaching

It has been observed that many of the ligands used in chromatography can become detached from the matrix and release into the purified product. Leaching of ligand is at its worst during the relatively severe conditions used for product elution, but ligand can be released at all phases of the chromatographic cycle. Released affinity-ligands can be difficult to remove since they may still be capable of binding tightly to the required protein when in their free state. Ligand leaching is also a particular problem if it occurs late in the purification sequence when little opportunity exists for further clean-up steps.

The ligands released may be the dyes used for pseudo-affinity separation or the linkers used to fix them to the column. Significant leaching of amine products from DEAE resins has also been reported (Boschetti, 1991). With this type of agent, the primary concern is one of possible chemical toxicity of the leached product even though the concentration of leachate may be only measured in nanograms per millilitre. The regulatory guidelines applied to recombinant protein products and monoclonal antibodies require the quantification of these agents and their removal by a validated purification procedure (Johansson, Hellberg and Wennburg, 1987; Lasch and Janowski, 1988).

The situation is worse when protein affinity ligands are used because leakage of ligand or its fragments is more likely because of the fragile nature of the molecule. Also the released protein may possess significant biological activity in its own right (as in the case of lectins or protein A) and may also be immunogenic. Again quantification of the level of contamination and the application of a validated purification procedure are required (Lucas et al, 1988).

Loss of ligand from the column can arise from several causes:

- release of trapped ligand that was not covalently bound.
- cleavage of the linkage between ligand and support.
- cleavage of the ligand itself (a problem particularly with protein ligands).
- dye stacking, a phenomenon in which several dye molecules associate tightly together but noncovalently.
- degradation of the support matrix.

All of these problems are being addressed in the design of new column materials. Very stable support matrices are now available, and improved linker chemistry has also contributed in a major way to reducing this problem.

Capacity of the purification to clear virus and residual DNA from cell derived proteins

As is discussed in the next chapter, much regulatory attention is currently directed to the cells used for product generation and their capacity to harbour biologically dangerous agents such as viruses or transforming DNA and to release them into the product stream.

The cells themselves are rigorously tested to determine what viruses may be present, and the downstream process has to be designed and validated so that it is capable of reducing the potential risks from these agents to acceptable levels. Specifically, when working from acceptable cell stocks it is necessary to reduce the final level of contaminating DNA to below 100 pg per dose of finished product and to show that the process is capable of removing or inactivating representative model viruses by a factor of 10^{12}.

Individual purification steps remove viruses with different efficiencies although most chromatographic methods in common use will fairly readily reduce virus load by a factor of 10^4. Three such steps in tandem would therefore achieve a 10^{12} reduction in titre so long as they all act independently. Ultrafiltration steps with membranes of appropriate porosity can also be very effective at clearing virus (Sekhri et al, 1992). In addition, many of the buffers employed during purification, particularly low-pH buffers, can have a major inactivating effect on virus.

Similarly, many purification steps, particularly ion-exchange procedures achieve major reductions in the concentration of residual DNA.

Validation of these steps is a critical aspect of downstream process development which requires early attention. Because the levels of both DNA and virus involved in a real production run would be below the limits of reliable detection, it is usual practice to validate procedures using 'spiking techniques', that is, adding virus or DNA deliberately into validation assays and demonstrating that the required clearance does indeed occur. Because such assays involve the use of DNA samples and live virus in quantities which would be difficult to obtain for a full-scale trial and because the equipment used would be potentially irreversibly contaminated, it is practice to perform the validations on a scaled-down version of the real process which accurately mimics the situation in a real purification run (Walter, Werz and Berthold, 1992). All of these approaches are discussed in more detail in Chapter 6.

It is interesting to note that satisfactory purification of the required protein could in some cases be achieved in fewer steps than are required to obtain the required virus clearance factor. Genentech is reputed to have developed an abridged purification procedure for tPA which eliminated some of the established purification steps. They were, however, reluctant to implement changes on this basis because they wished to retain the accepted virus clearance margin established with the old process. In this case the economic aspects of process improvements took second place to the established safety margin of a possibly less efficient protein purification.

Contamination with adventitious agents from protein affinity ligands

Obviously, reagents such as monoclonal antibodies used as purification aids carry the same potential risks regarding virus or other biological contaminants as does the protein product which is being purified. The regulatory guidelines are quite clear that monoclonal antibodies used in this way must be fully characterized. FDA regulation 21CFR points 200–299 define the principles of GMP to be applied to the manufacture of protein ligands before they can be included in systems for the purification of biopharmaceuticals.

Recent improvements in chromatographic materials for downstream processing

It is beyond the scope of this article to discuss in detail the advances which are now occurring in downstream processing materials. However, the following list serves to summarize some recent developments.

- Use of macroporous, rigid matrices permitting high capacity, rapid flow and high resolution.
- Development of purification systems in which ligands are attached directly to membrane structures. Examples include Zeta-Prep (LKB Productur), Affi-Sep™ (Anachem Ltd), Memsep ™ (Dominick Hunter Filters Ltd), Sartobind S (Sartorius) which are all ion-exchange membranes. Affinity purification membranes are also being developed.
- Development of synthetic ligands with high specificity for particular proteins such as Avid AL (Unisyn) which has similar performance to protein A and protein G in gamma globulin purification (Khatter, Matson and Ngo, 1991).
- Matrices and membranes resistant to much higher levels of NaOH (up to 1 M), acid treatment and autoclaving without degradation of the matrix or ligand leaching (e.g. Affi-Prep from Bio-Rad; Avid AL from Unisyn and many others).
- Radial flow-gel cartridges for increased flow without compaction.
- Multigram HPLC purification of proteins.

These developments have the effect of improving column separation efficiency, improving safety due to better CIP possibilities, reducing ligand leakages, and maintaining column efficiency after multiple cycles. Benefits are thus accrued in increased process robustness, reduced need for extra polishing steps and correspondingly reduced need for additional quality-control steps.

Characterization of recombinant protein products

To be acceptable as drugs, recombinant proteins and monoclonal antibodies must satisfy the same quality, safety and efficacy criteria as other pharmaceutical products. Precise characterization of the product and its batch to batch consistency are critical elements in satisfying these criteria. The objective of such analysis is to show that the biopharmaceutical product is pure or at least that any contaminants are identified and consistent, and that it authentically reproduces the naturally occurring molecule. However, the complexity of protein molecules and of the biological production systems used for their synthesis mean that this is not a simple objective to achieve.

A major analytical challenge with protein products is that protein molecules contain many possibilities for small variations in structure which may be very difficult to detect physicochemically, but which could affect the protein's biological and immunological properties and hence compromise the product's safety and efficacy. Uncontrolled or undetected introduction of such variations would mean that the required batch to batch consistency would not be achieved.

Origins of variant forms of recombinant proteins

Some possible sources of heterogeneity in recombinant protein preparations are summarized in Table 5.4. Of these possibilities, errors in the genetic construction is unlikely to be a problem because of the precision with which the construction can be verified and because the genetic stability of the production cell line will have been exhaustively characterized by analysis of the cell bank at various population doubling numbers as discussed in Chapter 6. However, mutations are not unknown, and it can be very difficult to detect a single amino acid change within a protein by physicochemical methods.

Post-translational modifications such as those listed in Table 5.4 can also have major effects on the functional competence and safety of the biopharmaceutical. As we have seen, control of these modifications during cell culture is much less rigid than that of the transcription of the required gene and can be influenced by a variety of environmental factors. Accurate analysis of product in terms of post-translational modifications is thus essential but remains one of the main difficulties in

Table 5.4 *Potential sources of modifications of recombinant protein structure compared with the equivalent natural protein*

Modification	Possible Cause
Variation in protein sequence	Errors/mutations in structural gene.
Post-translational modifications	Glycosylation γ-carboxylation of glutamic acid β-hydroxylation of aspartic acid Phosphorylations and sulphatations Amidation
Possible process-induced modifications	Deamidation of glutamine and asparagine Oxidation of sensitive residues (Met,Trp, Cys-SH) Scrambling of disuphide bonds Generation of 'ragged ends' by protease action, particularly at C-terminus Blocked terminal or ε-amino groups Amidation of C-terminus Isomerization of some amino acids

protein characterization because of the subtle nature of the changes that may occur.

In particular, the accurate characterization of the complex oligosaccharide structures added during protein glycosylation is a major problem which is currently under intense investigation (Parekh et al, 1989) and has formed the basis of a dedicated bioanalytical industry. This is because of the demonstrated sensitivity of the glycosylation process to culture conditions (Goochee and Monica, 1990) and its established influence on biological efficacy (Dube et al, 1988; Gribben et al, 1990). Process-induced modifications such as those listed in Table 5.4 can also be difficult to detect and identify and can have major impact on the acceptability of the product for drug use.

The complexity of protein molecules is such that no single characterization step can combine the sensitivity and the resolution necessary to define the molecule completely and accurately. The approach usually employed is to combine the results of a number of different tests based on different physicochemical criteria to generate a purity and characterization profile for the product. The tests used fall broadly into four categories:

1. Biological tests designed to confirm that the required biological potency and specificity are within the prescribed limits for the preparation
2. Tests designed to detect specific, potentially hazardous contaminants that may be present at levels too low for detection by physicochemical methods. Typically, this requires detection at the parts per million level and may involve the use of specific

antibodies for the detection of a suspected protein contaminant such as residual host-cell proteins or leached affinity ligands, pyrogenicity testing to detect and quantify endotoxin, hybridization or other DNA capture tests for residual DNA estimation, and sterility tests to demonstrate the absence of adventitious biological agents.

3. Tests of purity and identity of the recombinant protein itself. Because no single physicochemical property is unique to any specific protein, a battery of tests based on properties of the molecule are applied. These include analyses on the basis of charge (electrophoresis, ion-exchange, isoelectric focussing), molecular size (SDS-PAGE, size exclusion chromatography, mass spectrometry) and hydrophobicity (reverse phase chromatography). Immunological analysis and partial sequence analysis may also be performed.

   In some cases, the sensitivity of the test, particularly to minor changes in the interior of the molecule, can be improved by using peptide-mapping techniques after chemical or enzymic fragmentation of the protein.

   Most of the physicochemical techniques mentioned are incapable of detecting impurities below the 1–5% level. In addition, they are generally insensitive to small differences which exist between closely related complex molecules. Thus, minor structural variations in a recombinant protein preparation would be difficult to detect. The situation has changed radically with the introduction of a new generation of mass spectrometry instruments which can be used to measure directly the mass of intact recombinant proteins (or purified fragments derived from them) with a precision of the order of one mass unit. All of the modifications listed in Table 5.4 (except disulphide scrambling and amino acid isomerizations) involve a change in molecular mass and would therefore readily be detected by this technique (Monegier et al, 1990). Another advantage of this technology is the extreme rapidity (about 20 s) and economy (micrograms of sample) with which assays can be performed.

4. Verification of the physical properties of the protein (correct folding, tertiary structure etc.). Tests of this sort are much more difficult, expensive in material and time consuming, and are not performed on a routine basis. The techniques available for full structural characterization are X-ray crystallography and 2-D nuclear magnetic resonance. Useful but very limited information on protein tertiary structure can be obtained by techniques such as circular dichroism, Raman spectroscopy, intrinsic fluorescence and calorimetric studies. A summary chart of the range of the main analytical procedures available is given

Table 5.5 *Tests used to generate a purity/characterization profile for recombinant protein quality assurance*

Characterisation Parameter	**Methods**
Biological Properties	Potency Specific Activity Specificity Pharmacokinetics
Tests of homogeneity/identity	Electrophoresis Isoelectric focusing HPLC Peptide mapping N- and C-terminal sequencing Immunological analysis Mass spectrometry - fragment mapping - whole protein
Detection of Specific Contaminants	DNA hybridization test Immunological analysis Pyrogenicity tests Tests for adventitious agents
Conformational Analysis	X-ray crystallography 2-D nuclear magnetic resonance Circular dichroism Raman spectroscopy Differential scanning calorimetry Intrinsic fluorescence

above in Table 5.5. The general question of purity analysis of protein pharmaceuticals has been reviewed recently by Anicetti, Keyt and Hancock (1989).

Purity targets

Experience with biologicals has given rise to a consensus view that general levels of contamination by extraneous proteins should not exceed 100 ppm. The physicochemical chemical methods available, used to generate a characterization profile as indicated earlier, permit the demonstration of purity at about the 99% level. Obviously, the limits fixed for suspected contaminants that may adversely affect the safety or efficacy of the product can be set at lower levels. Specific, more sensitive tests are therefore required if the product is to be analyzed for such specified contaminants.

The full product characterization is only performed in its entirety once or a few times to demonstrate the authenticity of the product. For control of the quality of production batches, a limited panel of tests that have been validated as indicative of overall product quality are employed. In addition to the demonstration of the authenticity of the product as produced, it is also necessary to demonstrate its stability in the exact final formulation and container.

As has been emphasized at several points, it is the extraction, purification, formulation and subsequent testing regimes which ultimately bear the burden of demonstrating that the product is safe, effective, pure, stable and reproducible.

6 Regulatory Aspects of Using Cells as Bioreactors

General regulatory requirements for biopharmaceuticals

A full discussion of the general regulatory requirements for biopharmaceuticals is beyond the scope of this work but clear guidelines have been laid down in the FDA *Points to Consider* documents and in the European Community guidelines (Table 6.1), and these documents should be consulted for detailed information on the relevant test programmes.

In general the regulatory bodies call for

- full characterization of the starting materials: gene, genetic construction and host cell.
- full details of the establishment of the banks of production cells, of procedures used for their maintenance and of their stability.
- full details of the production process and the stability of the production cells during the process.
- full details of extraction, purification and characterization of the product.
- evidence of consistency of manufacture.

The guideline documents expand on these requirements and propose techniques that are acceptable and how these should be operated and the results interpreted.

Specific safety issues with animal cells

One of the risks of using cells for the production of biological products is that because they share many of the same biochemical capacities as the patients who will eventually receive the products they are also subject to some of the same pathological events including infection by viruses and the development of oncogenic changes. In principle, the agents causing such events could be transmitted from the production cells to the patients receiving the bioproduct. For this reason, in the early days of the use of cells as substrates for vaccine production, the FDA ruled that only normal cells derived from normal tissues could be used. In practice, this limited acceptable cell substrates to primary cell cultures which were rather poorly adapted to industrial use because new cell stocks must be produced from animal material for each production run, with the result that the cells used for different production batches could never be effectively standardized.

Later FDA guidelines permitted the use of diploid cell lines but this still limited the practical production of biopharmaceutical products because diploid cells have only a finite life span, have fastidious requirements for complex medium and do not generally grow in suspension.

For several years, largely theoretical concerns over the safety of tissue culture-derived products delayed the regulatory acceptance of continuous cell lines as substrates for production. Since then, progressively more compelling needs for vaccines and for purified protein products in quantities that can only be obtained from 'abnormal' or genetically modified cells have stimulated a rigorous and pragmatic re-assessment of the risks associated with manufacture from such cells. Continuous cell lines offer many advantages for bioproduct generation including infinite life span, relatively less complex medium requirements and the capability to grow in suspension culture. In many cases, the use of continuous cell lines may be the only way in which the required product can be produced economically.

New ground was broken in the early eighties when Wellcome received approval for the first human therapeutic produced in a continuous cell line. Wellcome used Namalwa, a transformed human lymphoma derived cell line, for the production of human interferon α (Pullen et al, 1985). Namalwa contains the EBV genome and is tumourigenic in animals. However, it was successfully argued and demonstrated experimentally that the biological risks associated with the cells could be eliminated from the purified biochemical product by a combination of rigorous purification procedures and denaturing conditions employed during extraction.

Following this precedent, the general regulatory position is now that there should be no a priori proscription of the use of any given cell type for the production of biologicals but rather that every situation should be examined on a case-by-case basis using risk-versus-benefit criteria. (The single exception to this would be cells derived from a host with disease of unknown aetiology which might reasonably be supposed to involve a transmissible agent.)

It is now therefore recognized that the use of continuous cell lines is a 'reality that cannot be avoided' if we are to develop cell-based production systems with yields that are acceptable for the commercial production of therapeutics. Once this principle is established, the main question becomes: how can the safety of products derived from such cells be assured? Current attention is accordingly focused on the realistic evaluation of the risks that might be associated with the particular cell used and the technology that needs to be applied to assure that these

Table 6.1 *Published international and national guidelines for the required regulatory approach to biopharmaceuticals*

FDA 'Points to Consider' Documents *	European Community Guidelines **
US	
Points to consider in the manufacture of monoclonal antibody products for human use, 1 June 1987.	Guidelines on the production and quality control of monoclonal antibodies of murine origin intended for use in man (Tibtech 6: G5-G8, 1988)
Points to consider in the production and testing of new drugs and biologicals produced by recombinant DNA technology, 10 April 1985.	Guidelines on the production and quality control of medicinal products derived from recombinant DNA technology (Tibtech 5: G1-G4, 1987)
Points to consider in the characterization of cell lines to produce biological products, 18 November 1987.	Guidelines on the production and quality control of cytokine products derived by modern biotechnology processes, (Tibtech 6: G9-G12, 1988)
Cytokine and growth factor pre-pivotal trial information package, 23 January 1990.	Validation of virus removal and inactivation procedures

* Available from the Centre for Biological Evaluation and Research of the FDA

** Printed as supplements in Tibtech

risks can be quantified in the cell line and during the process and thence to ensure that they are eliminated from the final product (Petricianni, 1988).

Cell bank system

Perhaps the most important advantage of using continuous cell lines is that their use facilitates consistency of manufacture by permitting the establishment of stable stocks of cells from which cells for successive production runs can be drawn when required. These cell stocks can be exhaustively characterized and subjected to intense quality assurance evaluation before being accepted for production. In particular, it is practice to test the cells after prolonged culture at population doubling numbers beyond that at which they would normally be used for production in the standard manufacturing process in order to assure that the cell line is genetically stable, that it maintains product yield and continues to produce authentic product, and that no latent virus or potentially oncogenic changes become activated during the cell line's usable lifetime.

This approach forms the basis of the cell banking system which is now universally employed. For manufacture, several different banks of cells are set up, beginning with the master cell bank (MCB) which is the basic reserve (typically 40–50 ampoules) of the original starting material for production cells. Accordingly, it is usual to deposit several ampoules of the MCB for storage in liquid nitrogen at several different locations to guard against accidental loss of this primary material.

After the completion of quality control tests, an ampoule from the MCB is used to set up a second larger bank called the working cell bank (WCB). The WCB is again quality controlled and one ampoule from the WCB is used to create the manufacturer's working cell bank (MWCB) which typically contains 2–300 ampoules depending on the anticipated demand. Ampoules from the MCWB are then taken out to initiate each production run. When the MCWB is exhausted, a new MWCB can be established from another ampoule removed from the WCB and so on throughout the production campaign (Figure 6.1).

Characterization of the cell line

A primary requirement for approval of a cell line is that it should be fully characterized so that the MCB cells constitute a properly defined seed stock. It is generally considered necessary to satisfy the characterization criteria listed in Table 6.2. Adequate characterization is critical since it provides an identity profile for the cells which allows verification that the cells do not alter over a range of population

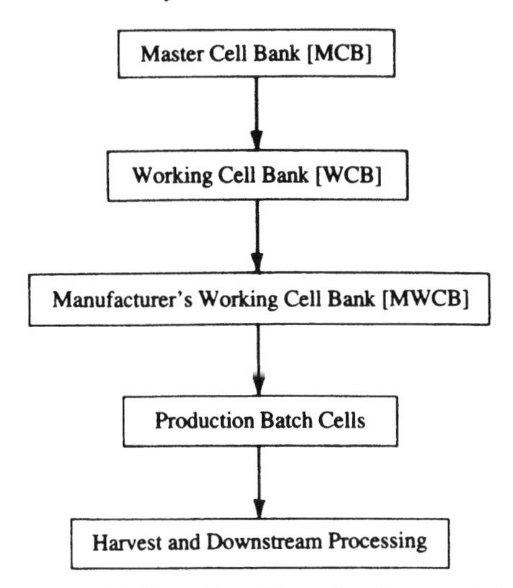

Figure 6.1 Flow sheet illustrating the use of the cell-bank (seed stock) system to permit standardization and full characterization of cells used for biopharmaceutical production. For security reasons, the master cell bank (typically up to 50 ampoules) is usually deposited at several different sites for storage in liquid nitrogen.

doublings which spans the doubling number at which the production process is operated. Evidence of this sort provides reassurance on the robustness of the manufacturing process. It is a regulatory requirement to demonstrate that the cell population at the end of the production process is identical to that in the cell bank and rigorous characterization is required to achieve this. This test of identity precludes the possibility of cross-contamination of the cells or of significant changes in the established characteristics of the cell line occurring during the production process.

Cell characterization procedures classically include analysis of distinguishing markers of the cell such as karyotypic analysis, isoenzyme profile and immunological features. One new analytical approach which deserves specific mention is the so called DNA finger printing or multilocus probe analysis technique.

This procedure was initially developed for forensic use (Jeffreys, Wilson, and Thein, 1985) but is finding increasing use in the characterization of cell stocks (Thacker, Webb, and Debenham, 1988). The method is based on restriction enzyme fragmentation of repetitive 'mini satellite' DNA to generate a pattern of fragments which is unique to the individual. This approach can readily detect cross-contamination between cell lines and can serve as a useful indication of genetic stability. Recent studies have shown that the technique can differentiate different hybridoma clones derived from a single fusion (Stacey et al, 1992). Multilocus DNA finger-printing gives an overview of the structure of

Table 6.2

Regulatory requirements for the characterization of cell lines
1. Documentation of the history and geneology of the cell line
2. Details of production and storage procedures used for the MCB and MWCB
3. Data which establishes the identity of the cell line:
Morphology
Karyology- particularly readily identifiable constant chromosomal markers
Isozyme analysis
DNA finger printing
Analysis of immunological markers
4. Growth characterization of the cell - details of the culture system to be used
5. Data which establishes the stability of the cell under the conditions used
6. Testing for viruses:
Classical virology
Tests for reverse transcriptase
Electron microscopic analysis
MAP or RAP tests if appropriate*
7. Absence of bacteria, fungi and mycoplasmas
8. Tumourigenicity testing
9. Tests for the expression of oncogenes

*Refer to guidelines (Table 6.1)

repetitive DNA surrounding the genes and so provides valuable information on cell line identity. However, it is not able to detect discrete changes in functional genes and for this purpose specific, single-locus probes would have to be developed.

Hazards associated with products from animal cells

The following perceived risks have been considered to be potentially associated with drugs derived from cultured animal cells.

Persistence of process-derived proteins

In principle, residual proteins derived from production cells or medium components which persist in the final product could provoke an immune response or could possibly cause transformation events. As was noted with some early vaccines, the presence of immunogens capable of cross-reacting immunologically with human tissues can cause severe clinical accidents in patients due to the stimulation of host allergic responses.

However, as discussed in the previous chapter, current high-resolution protein purification methods are able to produce protein preparations of extreme purity, and levels of persisting impurities may be below the limit of detection using the physicochemical methods available (Monegier et al, 1990). Detection of very low levels of impurity, which are normally impossible to see in the presence of a large excess of the required product, can be made possible by employing the mock-purification approach. This involves performing the standard purification procedure on supernatants derived from cells identical to the production cells except for the fact that they lack the gene for the product of interest. In this way, otherwise invisible contaminants can be identified and quantified.

Similarly, using antibodies against known potential impurities provides an extremely sensitive test for the detection of such impurities. This approach might, for instance, be applied to the detection of extremely low levels of residual bovine proteins in the final product if foetal bovine serum had been employed at some stage in the manufacturing process.

Overall protein purity levels in excess of 99% are routinely achieved, and experience has indicated that this level of purity is acceptable if it is also demonstrated that any potentially hazardous contaminants that are known or suspected to exist during the process have been eliminated from the final product.

Contamination by residual DNA

In the mid-eighties the persistence of DNA from continuous cell lines in biopharmaceutical preparations was considered to be a major risk factor potentially capable of producing pathogenic events in patients. The theoretical risks involved include uptake and expression of viral genes, cell transformation by activated oncogenes and the insertion of exogenous sequences into critical control regions of the genome with subsequent alterations in the expression of certain genes.

In 1984, 10 pg of DNA per dose of drug was recommended by the FDA as the maximum permissible dose in drugs derived from cultured animal cells, and this figure became accepted as an industry norm. The 10 pg level was based on the practical limits of DNA detection at the time and on theoretical analysis of the risk of transformation by biologically active DNA in the product. Assuming that, in a worst case, every production cell contained one activated oncogene and that, from the known transforming efficiency of purified cellular oncogene DNA in vitro, about 100 pg of oncogene was required to trigger a transformation event, it was calculated that the 10 pg limit would be below the minimum effective transforming dose for cellular oncogene DNA by a factor of about 10^8 (Petricianni, 1985). Regulations presented in the FDA *Points to Consider* documents and the European Community Guidelines (Table 6.1) require that the test procedures capable of detecting DNA at 10 pg per dose or below should be applied to biopharmaceutical products. Limits have been relaxed in certain cases, and the current WHO recommendations fix the limit at <100 pg per dose of final product (*Proposed Requirements for Monoclonal Antibodies for Clinical Use in Humans,* Technical Report BS/91, 1657, World Health Organization, 1991).

The size of the residual DNA is also a consideration, DNA which is too small to contain complete open reading frames obviously presenting a lower risk that DNA of a size permitting the encoding of intact proteins. However, reliable size estimation of DNA present at the picogram level is problematical.

Relaxation of the limits of acceptable DNA contamination appear to be justified by demonstrations that massive amounts of intact DNA from continuous cell lines can be injected into animals of the same species without any detectable oncogenic events or other harmful effects (Pulladino et al, 1987). It appears that naked DNA injected in this way is very rapidly cleared by nucleases and eliminated.

However, two recently published observations show that the risks of exposure to both activated oncogene DNA and to viral DNA are more than theoretical. Thus, Burns et al (1991) showed that mouse endothelial cells could be transformed in vivo by direct application of plasmid DNA containing the T24 H-ras oncogene to abraded mouse skin. In addition, Letvin et al injected adult macaques with single intramuscular doses of 200 μg of a recombinant phage containing SIV_{MAC} proviral DNA. Even though this phage preparation had been shown previously to be incapable of infecting macaque lymphocytes in vitro, three of the four animals became infected. Although the dose of DNA used in these experiments exceeded the level permitted in drug dosage forms by several orders of magnitude, it is clear that continued rigor and vigilance is essential in the area of contamination of biopharmaceuticals with residual DNA.

Elimination of residual DNA therefore remains a major objective in the design of purification processes, and sensitive assays are applied at each purification step to determine clearance factors for DNA. When DNA levels are too low for reliable measurement, clearance at the different purification steps is validated by spiking experiments in which radio-labelled DNA is deliberately introduced into the product stream and its clearance by a given purification step is demonstrated. In this way, it can be shown definitively that DNA levels can be reduced to acceptable limits (Builder et al, 1989).

Estimation of residual DNA

The most widely used assay for residual DNA is hybridization analysis in which DNA from the test article is quantitatively extracted, bound to a membrane and hybridized using species-specific probes. Spike recovery methods are used to ensure that the DNA is efficiently extracted from the sample, and the DNA signal from the test article is compared with that from DNA standards incorporated as controls in the hybridization run (Chou and Merigan, 1983; Per et al, 1989; Smith et al, 1992).

An alternative to hybridization assays, called the Threshold Total DNA assay has recently been introduced by the Molecular Devices Corporation. The Molecular Devices Threshold™ system is based on DNA capture by very avid but low sequence specificity DNA-binding proteins fixed to a membrane. The binding of DNA modulates the activity of an enzyme–anti-DNA antibody complex which also binds to the membrane (via the attached DNA). The activity of the enzyme then produces changes in the surface potential on an electronic chip (using a Light Addressable Potentiometric Sensor or LAPS system) which produces a signal proportional to the concentration of DNA present ('Threshold Total DNA Assay' Molecular Devices Corporation Publication 0120–0510A—1988, Molecular Devices Corporation, Menlo Park, California, USA).

The Molecular Devices system should provide simpler and more rapid estimation of total residual DNA in biologicals and has been shown to have similar sensitivity and reproducibility to the hybridization methods when carefully validated and standardized for analysis of a specific product (McNabb, Rupp and Tedesco, 1989; Smith et al, 1992).

Detection of specific DNA sequences

The methods discussed above measure total contamination with residual DNA. As will be discussed in the next section, specific DNA sequences can be detected by using specific probes in hybridization assays or can be detected with extreme sensitivity by use of the polymerase chain reaction (PCR).

Viral contamination of animal cell derived pharmaceuticals

The possibility of contamination of cell-derived (or tissue-derived) drugs by virus is a major concern. Past experience has shown that unknown

Table 6.3 *Possible origins of viral contamination of biopharmaceutical products*

Origin of Virus	Probable Virus Types Involved	Control Measures
Viruses endogenous to the cell line used	Retrovirus Herpesvirus	Characterization of cell banks Validation of purification process
Medium and process constituents (serum, trypsin, affinity ligands, etc.)	Pestivirus (e.g. BVD) Reovirus	Rigorous QC of medium components Avoid animal products in medium
Process contaminations due to the failure of GMP	Parainfluenza virus Reovirus	Rigorous control of antiseptic production under validated GMP conditions

or unidentified viral agents have been administered to humans as contaminants of vaccines and drugs. The most widespread incident involved the unknowing administration to hundreds of millions of subjects of killed polio vaccine which contained infectious SV40 virus which had survived the polio virus inactivation process. Fortunately, in this case there was no evidence of harm suffered by the recipients of the SV40 which replicates only very poorly in humans (Shah and Nathanson, 1976).

As is now well known, a far more tragic outcome resulted from the unsuspected contamination by HIV of human plasma used for the production of antihaemophilic factors. An important point in both of these cases, and a point which now forms one of the cornerstones of procedures for the exclusion of viruses from biopharmaceuticals, is that the contaminating virus was not recognized or identified until some years after the product had been in use in patients. The problem of unknown contaminants is thus central to the assurance of virological safety and this has been underlined again recently with the discovery of the bovine spongiform encephalopathy (BSE) transmissible agent in bovines in the United Kingdom (Collee, 1991).

Possible sources of viral contamination

Contamination of cell-derived products by viruses can arise from several origins. Viruses may be endogenous to the cell line or may enter as adventitious agents, either from the reagents used during production (such as animal serum used in cell culture or antibody preparations used in immuno-affinity chromatography) or by contamination via the human operators involved in the process (Table 6.3). Measures to combat all of these possibilities are inherent in the Good Manufacturing Practice (GMP) procedures adopted for the production of biopharmaceuticals.

Exogenous viruses borne by reagents are excluded from the production process by rigorous quality control of the reagents used. As discussed in Chapter 3, in the case of animal sera this will involve extensive virological testing and also the requirement to use serum originating in countries considered to be free from particular named viruses.

During the process, viruses (and other contaminants) are excluded by the use of validated aseptic production procedures and by the use of production plant and installations of appropriate quality for the manufacturing process.

The presence of endogenous virus within the cells is one of the issues addressed by the cell banking system. The regulatory guidelines available (Table 6.1) specify the methods which should be applied to search cell banks exhaustively for the presence of viruses. The cell bank system permits analysis over many population doublings to maximize the possibility of detection of viruses that may become amplified or induced during prolonged replication of the cells. Chemical induction of possible latent viruses must also be tested.

The methods used cover the whole range of virus detection techniques including electron microscopy, inoculation into different cell types capable of detecting a wide range of human, murine and bovine viruses, inoculation into susceptible animals, reverse transcriptase assay for the detection of retroviruses and specific antibody test procedures (for details refer to the documents listed in Table 6.1). The guidelines also provide indication of viruses which should be given special consideration as potential contaminants in different specific types of cell, and those which, because of their known association with disease in man, are not allowable as contaminants in cell lines intended for use as substrates for bioproduct generation.

Despite these specific exclusions, many of the continuous cell lines which are employed in manufacture do contain certain viruses. As in the case of the contaminating DNA discussed earlier, removal of these viral contaminants is then based on the validated procedures for their elimination or inactivation, and the subsequent testing of the product. Knowledge of which viruses may be present in the MCB provides guidance as to which purification procedures should be the most effective and what specific tests should be applied to ensure that potential viral contaminants have been removed. Testing at different population-doubling numbers also provides information on viruses that might become induced during the production process. This is particularly important for cells destined for long-term use in continuous production processes.

Contamination by viruses of any kind obviously poses risks. Because of their particular properties of latency and their ability to cross species

Table 6.4 *Continuous cell lines currently used in production of products for clinical use*

Cell	Virus Elements Identified	Tumourigenicity	Product
Namalwa	EBV	+	α interferon
CHO	Retrovirus	+	Various Proteins
Hybridomas	Retrovirus	+	Monoclonal Abs

barriers, retroviruses and herpesviruses are of particular concern. Several of the cell lines currently used in production may carry elements of these viruses (Table 6.4).

Contamination by retroviruses

Retroviruses have several properties which make them particularly worrisome as potential contaminants of biological products:

1. Retroviruses are associated with oncogenic and immunosuppressive disease.
2. Endogenous retroviruses are inherited as DNA provirus integrated into the chromosomes of the germ line.
3. Exogenous retroviruses can be acquired by infection.
4. Retrovirus infections frequently do not produce cytopathic effects and remain as latent infections. As such they are not recognizable by standard virus detection techniques.
5. Retrovirus host range and virulence may become altered by recombination between retrovirus and cellular sequences such as provirus and oncogenes.
6. Host range can also be modified by the formation of pseudotypes in which the genome of one virus becomes packaged in the envelope of another.

Several of the cells which are widely used for the production of biopharmaceuticals contain retroviral elements (Table 6.4). For instance, electron microscopy reveals low levels of retrovirus-like particles in all CHO cells. Both intracytoplasmic A-type particles and budding C-type particles are observed (Anderson et al, 1990). Careful analysis of these particles has indicated that they are noninfections particles derived from endogenous retrovirus-like elements in the CHO germ line, which are defective in encoding the endonuclease which is essential for replication (Anderson et al, 1991). Despite the defective nature of the particles, the possibility of the generation of a competent virus by recombination as discussed above is always present.

Mouse hybridomas also frequently contain retroviruses of several classes. Ecotropic viruses are able to infect rat or mouse cells, and xenotropic retroviruses can infect a wider cell range including humans. Recombination between ecotropic and xenotropic viruses to produce polytropic viruses that are also able to replicate in a wide range of hosts has been widely observed.

A recent electron microscopic survey of ten CHO cell lines and 10 hybridoma lines, all in current use for manufacture, showed that every one of the lines tested was capable of expressing both C-type and intracisternal A-type retroviruses, although the level of expression varied by several orders of magnitude. In the case of two of the CHO cell lines, C-type particles were seen only after induction with bromodeoxyuridine (Liptrot and Gull, 1992).

Herpesvirus contamination

Herpesviruses are also a particular concern due to their widespread presence in animal and human populations, their association with oncogenic and systemic disease and their capacity to persist as latent infections without production of cytopathic effect and hence to evade detection by classical virological tests. EBV, a lymphomogenic herpesvirus, is present in Namalwa cells and many other transformed lymphoid cell lines (Table 6.4).

Recently a new mouse herpesvirus of the same class as EBV (the gamma-herpes viruses) has been identified and shown to occur in some murine hybridoma lines. It is recommended that this virus is also specifically tested for in murine monoclonal antibody preparations.

Approaches to the elimination of viruses from biopharmaceuticals

The discussion so far indicates the type of hazards that can be associated with viral contamination of drugs and shows that, for two potentially dangerous types of virus, the kinds of continuous cell lines used for bioproduct generation may be contaminated and potentially capable of releasing infectious viruses. In addition to these identifiable contaminants, the possibility of contamination by an unknown infectious agent is always present. Processes for the elimination of virus must therefore achieve demonstrable clearance of known viruses and provide the best assurance possible that any unknown virus would also be effectively cleared.

If testing shows that virus is present in the cell line or in the bulk harvest, it is necessary to show that the extraction and purification procedures applied to yield the final pure product can also effectively re-

Table 6.5 *Criteria for the selection of viruses for the validation of virus clearance in downstream processing*

Representiveness	Practicality
* Viruses used must represent a wide range of agents with respect to: Size Sensitivity to inactivation Envelope structure Nucleic acid type and structure * Viruses must be included that are relevant to the production cell line (e.g. retroviruses for hybridomas)	* Viruses must be readily available at high titre (to permit measurement of large clearance factors) * They should be detectable at high sensitivity * Viruses used should be easily and rapidly quantifiable

move virus to a degree that eliminates all reasonable risk. As with the reduction of residual DNA to acceptable limits, this is achieved by the application of a validated downstream processing sequence. Virus clearance should ideally be achieved by a combination of both purification and virus inactivation steps.

Validation of virus elimination

Since the virus levels likely to be encountered during the actual production process are at or below the limits of reliable detection, it is necessary, as for residual DNA clearance validation, to use spiking techniques, this time with infectious virus, to demonstrate clearance at each purification step. Setting up this type of validation experiment involves several practical problems.

Selection of model viruses

Obviously, viruses vary greatly in properties such as size, resistance to inactivation, type of envelope, type and structure of their genome. When information is available (from studies on the MCB for example) to indicate what contaminating viruses might be present, the ideal is to use that specific virus for spiking experiments (for instance, murine retrovirus particles in the case of murine hybridoma cell lines). However, it is necessary to be able to spike with a virus that can be grown to high titre (in order to measure large clearance factors) and which is easy to assay. It may therefore be necessary practically to use a virus closely related to the contaminating virus, such as murine xenotypic leukaemia virus in the case of murine retrovirus.

To cover the risk of unknown viruses, it is necessary to validate the clearance of representative examples of several different virus types spanning a range of physicochemical characteristics (Table 6.5). Lists

Table 6.6 *Examples of viruses which have been advocated for use as model viruses in virus clearance validation studies (Horaud, 1991)*

Virus	Virus Family	Size (nm)	Genome	Envelope	Sensitivity to inactivation	Natural Host
Polio	Picornavirus	25	RNA	No	Medium	Human
Reovirus	Reovirus	60	RNA	No	Low	Various
SV40	Papovavirus	45	DNA	No	Low	Simian
Murine Leukaemia	Retrovirus	80	RNA	Yes	High	Murine
HIV	Retrovirus	90	RNA	Yes	High	Human
VSV	Rhabdovirus	80	RNA	Yes	High	Bovine
Parainfluenza	Paramyxovirus	150	RNA	Yes	High	Various
Pseudorabies	Herpesvirus	120	DNA	Yes	Medium	Porcine
Vaccinia	Poxvirus	600	DNA	Yes	Low	?
IBR (1)	Herpesvirus	100	DNA	Yes	Medium	Bovine
Adeno (1)	Adenovirus	70	DNA	No	Medium	Murine
Canine parvovirus	Parvovirus	22	DNA	No	Low	Canine

(1) Walton et al, 1992
(2) Harbour et al, 1992

of candidate model viruses have been published (Table 6.6), and a consensus is emerging of the most useful virus types (Horaud, 1991).

Scale-down

Ideally, validation should be performed in equipment which is identical to the actual equipment used in the manufacturing process. Clearly this is impractical, both from the point of view of contaminating expensive process equipment and of the massive amounts of test virus which would need to be produced for full-scale spiking studies. In practice, a small-scale version of the production equipment is used, designed so that the critical process parameters are held constant so the results obtained are representative of those achievable in the full-size plant.

Specifically this includes using the same chromatographic supports and ultrafiltration membranes. In the case of columns, bed height and linear flow rate must be identical between the model and the full-size plant. Membrane-based separation steps should have the same flow of fluid per unit area.

Calculation of clearance factors

Clearance at each process step should be evaluated separately. For two steps which use different criteria for separation (e.g. ultrafiltration and ion-exchange) the steps can be considered to be multiplicative, that is, the total clearance is equal to the sum of the log of the clearance factor at each step. However, when successive steps are evaluated separately it is important to ensure that they act in a truly independent fashion in the overall process or the real clearance obtained may be less than the sum of the log clearances for each step (Lees and Onions, 1991).

It should also be noted that when clearance involves the virus binding to the chromatographic matrix, it is possible that virus can accumulate on the matrix and be released during subsequent purification runs, thus contaminating later batches of the product. Properly validated column cleaning procedures must, therefore, also be incorporated into the final process to ensure that all contaminants are eliminated from chromatographic supports between runs.

Processing steps used for virus removal

Selective binding of product in affinity chromatography usually results in very efficient virus clearance. However, affinity columns using protein ligands are very expensive and may themselves add to the regulatory problem if, as in the case of monoclonal antibodies, they are also derived from animal material. Other chromatographic methods such as ion-exchange chromatography and hydrophobic chromatography tend to give lower clearance factors, but even so, clearance factors of 4 logs per step are realistically obtainable.

Filtration using appropriate size cut-off membrane also gives very useful reductions in virus titre, although this technique has limitations when it is required to separate very large proteins such as IgM from small viruses since both particles are of similar diameter.

Virus inactivation by pH change, solvents, heat or chaotropic agents is also of great usefulness if the conditions required to inactivate the virus can be tolerated by the required protein. The sensitivity of virus types to these inactivating agents varies enormously. In addition, other factors such as protein content or salt content of the virus suspension can profoundly influence the kinetics of virus inactivation. In particular, the formation of clumps of virus and aggregation of virus with protein are both factors which can result in greatly enhanced resistance to virus inactivation.

Generally, it is required to demonstrate a cumulative virus clearance during the process of 12 logs or more, and this is fairly readily achieved using a typical purification sequence involving several chromatographic steps, filtration steps and stages which include virus inactivation conditions.

Testing final products for virus

As detailed in the various guideline documents (Table 6.1), the bulk final processed product should be inoculated into cell cultures capable of detecting a wide range of virus types. Specific tests such as the XC plaque assay or the S^+L^- focus assay should also be performed. In addition to these infectivity tests, the use of PCR techniques for the detection of retroviruses and herpesviruses has also been proposed (Onions and Lees, 1991). In particular cases in which analysis of the MCB has indicated the possibility of contamination by other viruses, these should also be specifically tested for in the final product using specific hybridization or PCR techniques.

It should be noted that no test procedures currently exist for the practical detection of contamination of products by agents of the prion type which are implicated in the spongiform encephalopathies. Protection against these is entirely based on their presumed absence from input materials such as serum and their effective clearance by the downstream processing employed (Darling et al, 1992).

A new fully validated process for the inactivation of BSE in serum and in purified proteins has recently been announced by Hamosan GesmbH (Graz, Austria).

Conclusions

The persistence of potentially harmful contaminants in products derived from animal cells is prevented by the application of validated purification procedures for the elimination of residual DNA, contaminating proteins and viruses. Additionally, in the case of viruses, there is also intensive screening of the starting cellular material to ensure that specified proscribed viruses are not present in the primary stocks and that any viruses that may be present are identified.

The approach taken to verify that the final product is uncontaminated by these agents is a 'belt and braces' approach. Identified suspected contaminants are specifically tested for by immunological analysis in the case of proteins and by techniques such as PCR in the case of viruses and DNA fragments. This is in addition to the general tests of purity and for the presence of adventitious agents that is applied to each production lot.

Possible unknown risk factors are minimized by the application of validated processing techniques that are shown by spiking studies to be effective in clearing DNA and a range of different virus types from the product by a numerical factor which reduces any theoretical risks to acceptable levels.

The safety of products derived from animal cells is best assured by understanding the risks that may exist and evaluating them pragmatically, and by rigorously and intelligently applying the tests and validation procedures currently advocated by the international regulatory authorities.

7 Overview and Conclusions

Over the past ten to fifteen years, animal cell biotechnology as a means of producing pure protein products (as opposed to vaccines) has advanced from being an essentially research laboratory-based technology to being a fully fledged player in the biopharmaceutical manufacturing industry. An increasing proportion of products of major clinical importance are produced in animal cells, essentially because accurate and authentic post-translational processing is necessary for the effective functioning of many diagnostic and therapeutic proteins.

This rapid growth to maturity has been made possible by rapid advances in the wide range of technical areas which need to be brought together to achieve safe and cost effective product generation in animal cells. This book has sought to identify these key areas of advancement, to review present progress and to highlight remaining obstacles. The following points represent the major landmarks in the development of animal cells as usable bioreactors and indicate where future research emphasis might be placed to further improve quality and productivity.

The altered regulatory stance to animal cells used as bioreactors

Probably the key event in the resurgence of animal cells as bioreactors for the production of pharmaceuticals was the radical evolution of regulatory attitudes which permitted the use of continuous cell lines as substrates for production. This was made possible by, on the one hand, advances in knowledge of the risks that products from animal cells might pose to patients and a rational assessments of these risks relative to the benefits, and on the other hand, by the development and acceptance of technologies able to reduce these risks to acceptable levels. No type of cell is now formally proscribed a priori from use for the production of biopharmaceuticals provided that the cell is fully characterized and that properly validated procedures are applied to its use and to the extraction and purification of products derived from it. In practice, several high-performance cell types now have an established track record in pharmaceutical production (such as CHO cells, BHK 21 cells and various hybridomas, myelomas and lymphoblastoid cells), and processes based on these cells are likely to proceed more rapidly through regulatory examination than products derived from untried cell types.

Insect cell expression systems have also appeared on the scene and have an important role to play which has not so far been fully exploited. It is generally considered that insect cells may represent a desirable production substrate from the biological hazard point of view since they are unlikely to harbour any potential human pathogens. Although current experience suggests that glycosylation patterns in insect cells differ from those obtained in mammalian cells, this conclusion is based on analysis of products from only a very few insect cell types. Knowledge of the physiology of insect cells and how culture conditions may affect glycosylation and secretion of proteins is also at a very early stage. The range of vectors studied in insect cells remains very limited at present and this area also requires further study.

The relaxation of the previous regulations which limited production to strictly diploid cells opened the way to the use of cells of unlimited life span which can be cultivated on a large scale using new fermentor configurations that are not well adapted to the growth of diploid cells. It also opened the way to the use in production of newly created types of cells such as hybridomas and cells transfected with exogenous genes placed under the control of powerful promoters and enhancers capable of driving levels of protein expression unobtainable in normal diploid cells. For example, the maximum level of beta interferon obtainable in diploid fibroblasts was of the order of 4×10^4 u/ml compared with levels of 10^7 u/ml obtained in recombinant CHO cells, and yield of tPA was 0.2 μg/ml in the highest producing naturally occurring cells (Bowes melanoma) compared with 50 μg/ml in engineered myeloma or CHO cells.

Expression of heterologous proteins in animal cells

As discussed in Chapter 2, many systems have now been shown to be successful in obtaining high levels of transcription of heterologous genes in animal cells. Although there is still need for the development of more efficient transfection procedures, better amplification methods and methods for directing the required gene to the regions of the genome which are most efficiently transcribed, attention is shifting increasingly to the influence of post-translational events on the eventual secretion of recombinant proteins. This work is still in its infancy but several lines of investigation indicate that the processing and secretory mechanisms may be limiting in cells which produce very high levels of heterologous mRNA, and that limitations at this level may have effects on the yield and the authenticity of the product. Although some data are available regarding proteolytic processing steps and N-linked glycosylation, very little is known about the enzymes involved in other modifications such as gamma carboxylation, sulphatation, phosphorylation or O-linked glycosylation, all of which critically affect the functionality of some proteins. Once identified, the key enzymes involved in processing can be manipulated, either by reducing their level using anti-sense technology, or by engineering increased expression. This approach has already been attempted in a few cases (see Chapter 2) and will be more rationally applied as the mechanisms of processing and secretion control become better understood. Overall, knowledge of processing pathways and how they can be manipulated are topics which require in depth investigation in the future.

Yield and authenticity may also be affected by limitations in the availability of precursors for glycosylation and other processing steps. Such limitations can occur in high density cultures where nutrient supply can be suboptimal, at least locally.

Large-scale growth of animal cells

Advances in this area have also been considerable. A wide range of different types of fermentor have now been operated successfully and the previously held dogmas of the fragility of animal cells and their nonsuitability for the application of mass-culture techniques have been largely laid to rest. Studies in various fermentor configurations have converged to show that oxygenation of cells may be the main limiting 'nutritional' factor, particularly in high-density cultures. Much attention is currently focused on how to deliver adequate oxygen supply without damage to the cells caused by foam generation when gas sparging is used. The use of microbubble sparging techniques and the physical separation of the cells from the oxygenation zone in the medium are both promising approaches to the resolution of this problem. The use of additives such as Pluronic F68 can also reduce the damaging effects of bubble bursting and disengagement.

High-density fermentors offer advantages by providing a high concentration of product in a relatively small volume. However, as discussed in Chapter 3, this may be offset to a greater or lesser degree by the difficulties of precise control of culture conditions in such fermentors.

The control and monitoring of cell growth, viability and product generation is one of the key areas requiring further study. No satisfactory way exists at present of following many of the important culture parameters in real time. This is important both for the reproducibility of production runs and for early detection of deviations occurring from the established process norms. Recent studies have shown that post-translational modifications of protein may be very sensitive to changes in the culture environment and therefore accurate process control can have a direct effect on product authenticity as well as on product yield.

The process control problem is being approached both by the design of new, near on-line monitoring instrumentation and by the development and validation of appropriate mathematical models (Chapter 3). However, much remains to be accomplished in both of these areas.

Tissue culture medium

Medium design has also advanced greatly over the last few years. Defined media are now available for most cell types, and it is not usually necessary to supplement cultures with animal serum. This avoids several major and well-documented problems which have been classically attributed to medium supplemented with material of animal origin including variability of yield due to uncontrollable variation of medium composition, risk of contamination with agents imported with the animal material, and excessive protein load in the product stream passing for downstream processing. In reality, many existing processes still use serum because the development of reliable serum-free medium post-dates the time at which these processes were filed for regulatory approval. However, it is very clear that any new processes now in planning should be designed to operate using serum-free or protein-free medium.

One phenomenon which requires further investigation is that of programmed cell death or apoptosis. It appears that, for some cells, continuous stimulation by specific growth factors via cell-surface receptors may be essential to prevent activation of a preprogrammed 'suicide' sequence. If this is the case in culture, it may not be possible to completely eliminate exogenous protein from the medium for these cells. However, autocrine production of the required factor may occur naturally or could be engineered into the cells.

Physiology of cultured cells

Although it is known that many external factors and reagents influence the productivity of cells, little is known regarding the mechanisms by which this occurs and optimization of such factors is still essentially empirical. In some cases, such as the various alternative pathways of glutamine utilization or the different enzymatic steps involved in glycosylation (Chapter 4), some data is available, and approaches to controlling the key reaction pathways in order to favour optimum product generation have been suggested. For the most part however, the cell remains a 'black box' with few inputs being monitored and their effects on the outputs being poorly understood. This is particularly true in the case of insect cells of which knowledge of the physiology of cultured cells is truly minimal at present.

Thus, cellular physiology is also an area where substantial further research is required. The relationship between the study of cellular physiology and the ability to monitor and model cell responses is clear, as is the relationship between these elements and the design of a fully optimized and defined medium. Integrated studies in these areas are likely to be particularly fruitful in tuning the cells for maximum production of authentic product.

Product purification and characterization

Advances in efficient protein purification methods and in the analytical procedures required to demonstrate the purity of the product and the absence of proscribed contaminants have also been central in permitting exploitation of the animal cell production systems now available. As discussed in Chapters 5 and 6, effective, validated purification is one of the main struts supporting the regulatory position which permits the use of continuous cell lines.

Protein chemistry techniques are also particularly important for the verification of product authenticity. The production of different glycoforms depending on subtle changes in cell culture conditions has recently been identified by the application of newly available analytical techniques. This is a problem which is currently receiving great attention and which may have to be resolved by better control of the glycosylation processes in the cells as discussed earlier. An alternative approach, now becoming feasible with recent advances in glycoprotein chemistry, might be the removal of incorrectly glycosylated product in the downstream processing sequence. Again the interactive nature of these facets of the process is apparent, the success of the downstream processing being dependent on upstream process design and vice versa.

An overview of the general state of the art for several of these key elements is given in Table 7.1.

Integrated process design

A recurrent theme when discussing the optimization of product generation in animal cells is the interactive nature of the different elements involved and the need to take account of this in the design of an integrated process. Too frequently in the past, high-yielding recombinant cells have been constructed and then much of the rest of process development has been concerned with resolving the problems imposed by this choice. Such difficulties can be minimized if the process is considered holistically from the beginning. In particular, the following points should be considered and evaluated at the earliest possible stage in process design.

Table 7.1 *General overview of current progress in key areas in the use of animal cells as bioreactors for manufacture of biopharmaceuticals*

Technical Area	Issues that are largely resolved	Issues still requiring resolution
Expression of heterologous protein	· High transcription levels are achieved in several systems	· Control of transcription · Adequate and correct post-translational processing and secretion · Better knowledge of insect cells
Medium design	· Serum-free media exist for many cell types · Completely protein-free media exist for some cells	· Optimization of protein-free media for specific cell types in specific production systems
Fermentor design	· A wide variety of fermentor configurations are established permitting large scale culture of most cell types	· More efficient oxygenation of cells without cell damage · Improved monitoring of cell conditions · Control of cell cultures based on the physiological state of the cells · Better homogeneity in high density fermentor configurations
Cell physiology	· Present knowledge is limited	· Control of differentiation and the switch between cell growth and product generation · Limitation of toxic product production · Ideal nutrient feed programmes · Role of physiological changes in post-translational processing
Downstream processing	· Very effective purification procedures exist · Very effective characterization procedures exist	· Development of more and better specific synthetic ligands · Improved stability of membranes and column support material for better CIP · Higher capacity, higher flux systems
Regulatory	· Approval framework for the use of high yield continuous cell lines · Adaption of the batch concept to incorporate production from continuously harvested culture systems	· Rationalization of some aspects of the required procedures e.g.: · is there a logical need to test hybridoma cells for tumourigenicity or to re-evaluate the virological status of CHO cells of known provenance? · Development of simpler validated tests for DNA, virus, etc. · Complete integration of continuous cell culture processes into the regulatory framework

1. Choice of production cell line

- Cell line must be acceptable from a regulatory viewpoint.
- Cell line must be capable of the appropriate post-translational modifications for the required product.
- It must be practical to grow the cell line on a scale and in culture conditions compatible with the anticipated production volume required.
- Cells for which there is considerable experience of growth on

a large scale and for which good serum or protein-free media exist would be a particularly attractive choice.

2. Genetic constructions used

- Yield must be optimized for the chosen product but note that this includes verifying the capacity of the cells used to produce accurately processed protein product at the transcription rates achieved.
- Consider the possibility of using genetic constructions which might aid subsequent recovery of product, for instance, use of an affinity tag to facilitate purification or specific site mutagenesis to eliminate a fragile cleavage site in the protein to prevent nicking during production and recovery.
- Stability of the genetic construction must be compatible with the number of generations for which the cell is required to be in production. This is a function both of the scale of production required and of the type of fermentation used (continuous or batch).
- Evaluate the effect of the genetic manipulations employed on other factors which influence the suitability of the cell for production use such as increased doubling time or fragilization of the cells. These factors may be important for culture scale-up, for instance, increased shear sensitivity may be incompatible with growth in a stirred tank fermentor.

3. Design of the cell culture operation

- Obtain the maximum biomass possible *consistent with optimal expression of product.*
- Production of the maximum biomass possible consistent with good product recovery, for example, avoid cell conditions or cell densities in which cell leakage or cell lysis become significant.
- Design medium to facilitate downstream processing. This involves restricting the addition of exogenous proteins to the minimum compatible with acceptable cell growth and product generation, and also with the stability of product secreted into the medium. Other process additives which must be removed later such as antifoams or antibiotics should also be avoided whenever possible.
- Although medium design should respect the previous points, and medium should be tailor-made for the production cells

as far as possible, the cost of the medium must be compatible with that of the final product.
- The fermentor configuration used must be matched to the growth characteristics and the culture demands of the chosen production cells and must be scalable to the final manufacturing level anticipated for the product.

4. Downstream processing

- Purification procedures must be designed to suit the nature of the product stream and also the scale of operation required.
- To preserve high yield, use the minimum number of purification steps compatible with obtaining acceptable product quality. Use high-resolution steps as early as possible to reduce the scale of unit operations quickly.
- As far as possible, adapt successive purification steps to the product stream from the preceding step to avoid the need to adjust solution conditions (for example, use a hydrophobic chromatography step in high salt to follow an ion-exchange step).
- Choose chromatographic supports that allow sufficient throughput of material to suit the required scale of production, which have a high capacity for the product and which can be efficiently cleaned in place without loss of performance to permit use for multiple production cycles.
- Include purification steps that give high virus clearance factors. Wherever possible, if the stability of the product allows, apply conditions that produce significant inactivation of virus.

5. Regulatory considerations

- As discussed in Chapter 6, each phase of the production process has to satisfy the relevant regulatory requirements. The parallelism that exists between the successive manufacturing steps and associated regulatory concerns is illustrated in Fig. 7.1. One very important point to keep in mind is that once process details have been filed with the regulatory agencies, it will be expensive and time-consuming (although perfectly possible) to incorporate significant changes in methodology (Durfor and Scribner, 1991). It is thus critical to ensure that the process filed is capable of supplying the required quantity of product as the market develops.

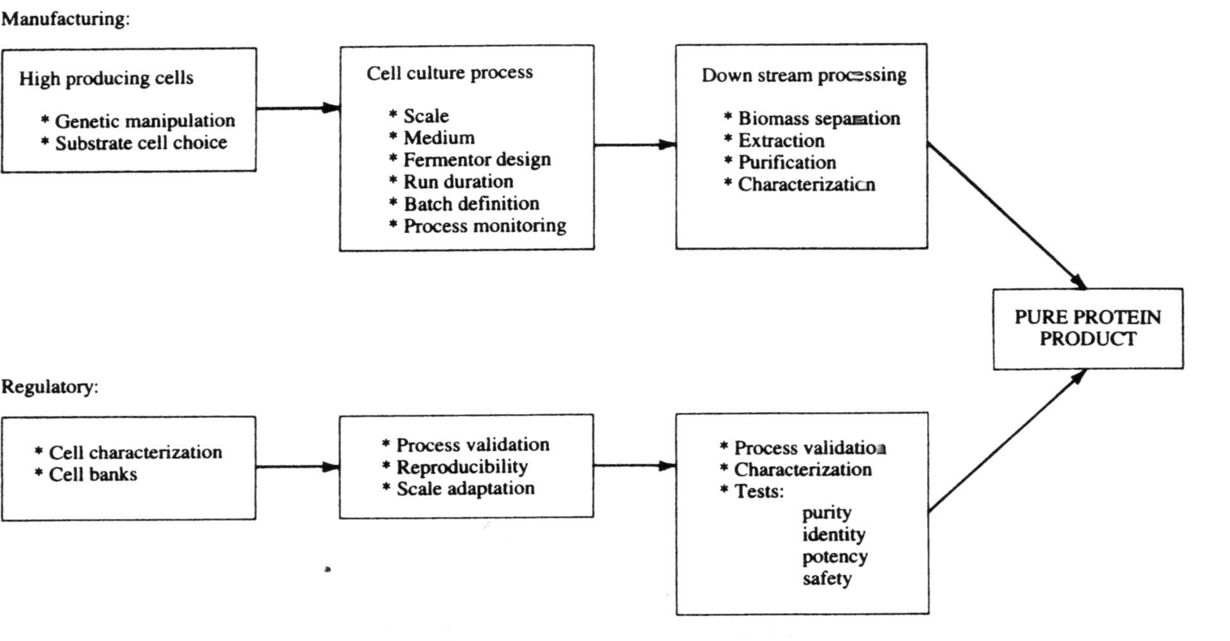

Figure 7.1 Interaction of the manufacturing and the regulatory aspects of process development for the production of biopharmaceuticals from animal cells

6. Commercial aspects

- The need for speed of development may compromise the possibility of fully optimizing all aspects of the process, for example it may be more time-effective to use an existing, well-understood process rather than investing time to optimize an untried process which would ultimately give better yield. The earlier the new process can be evaluated in a fully integrated fashion, the less there will be the necessity to compromise in this way.
- As in all manufacturing processes, overall cost of product development has to be balanced against projected selling price and anticipated sales volume.

Overview

The overall aim is to develop a robust process capable of consistently producing good yields of product of reproducible quality. Production of biopharmaceuticals in animal cells is a complex process that requires the simultaneous optimization of a series of interactive operations embracing both manufacturing and regulatory requirements as summarized in Fig. 7.1.

Attention to the different facets of the use of animal cells for the production of proteins which have been discussed in this book will permit the development of processes that are practical, acceptable from a regulatory viewpoint and economically viable. In these conditions, the use of animal cells for the manufacture of pharmaceutical proteins will continue to expand and to make full use of the synthetic versatility and fidelity of animal cells as bioreactors.

References

Al-Rubeai, M., Chalder, S. & Emery, A.N. (1991), Dynamic assay of synthetic activities in single hybridoma cells. In *Production of biologicals from animal cells in culture*, eds. Spier, R.E., Griffiths, J.B. & Meignier, B. London: Butterworth-Heinemann, pp. 587–592.

Al-Rubeai, M., Kloppinger, M., Fertig, G., Emery, A.N. & Miltenburger, H.G. (1992), Monitoring of biosynthetic and metabolic activity in animal cell culture using flow cytometric methods. In *Animal cell technology: developments, processes and products*, eds. Spier, R. E., Griffiths, J.B. & MacDonald, C. London: Butterworth-Heinemann, pp. 301–6.

Almgren, J., Nilsson, C., Petersson, A-C. & Nillson, K. (1991), Cultisphermacroporous gelatin microcarrier-new applications. In *Production of biologicals from animal cells in culture*, eds. Spier, R.E., Griffiths, J.B. & Meignier, B. London: Butterworth-Heinemann, pp. 434–8.

Amann, E., Abel, K., Grundmann, U., Okazaki, H. & Kuepper, H.A. (1988), Synthesis of human factor XIIIa in bacterial cells. *Behring Inst. Mitt. 82*: 35–42.

Anderson, K.P., Lie, Y.S., Low, S.R., Williams, E.H., Nguyen, T.P. & Wurm, F.M. (1990), Presence and transcription of intra-cisternal A particle-related sequences in CHO cells. *J. Virol. 64*: 2021–32.

Anderson, K.P., Low, M-A.L., Lie, Y.S., Lazar, R., Keller, G. & Dinowitz, M. (1991), Defective endogenous retrovirus-like particles of Chinese hamster ovary cells. In *Production of biologicals from animal cells in culture*, eds. Spier, R.E., Griffiths, J.B. & Meignier, B. London: Butterworth-Heinemann, pp. 39–44.

Anderson K.P., Low, M.L., Lie, Y.S., Keller, G.-A., & Dinowitz, M. (1991) Endogenous origin of defective retrovirus-like particles from a recombinant Chinese hamster ovary cell line. *Virology 181*: 305–11.

Anicetti, V.R., Keyt, B.A. & Hancock, W.S. (1989), Purity analysis of protein pharmaceuticals produced by recombinant DNA technology. *Tibtech 7*: 342–9.

Archer, R. & Wood, L. (1992), Production tissue culture by robots. In *Animal cell technology: developments, processes and products*, eds. Spier, R.E., Griffiths, J.B. & MacDonald, C. London: Butterworth-Heinemann, pp. 403–7.

Armiger, W.B., Forro, J.F., Montalvo, L.M. & Lee, J.F. (1986), The interpretation of on-line process measurements of intracellular NADH in fermentation processes. *Chem. Eng. Commun. 45*: 197–199.

Augenstein, D.C., Sinskey, A.J. & Wang, D.I.C. (1971), Effect of shear on the death of two strains of mammalian tissue cells. *Biotech. Bioeng. 13*:409–18.

Avgerinos, G.C., Drapeau, D., Socolow, J.S., Mao, J., Hsiao, K. & Broeze, R.J. (1990), Spin filter perfusion system for high density cell culture: production of recombinant urinary type plasminogen activator in CHO cells. *Bio/Technology 8*: 54–58.

Barford, J. & Harbour, C. (1991), A computer simulation of the kinetics and energetics of hybridoma cell growth and antibody production. In *Production of biologicals from animal cells in culture*, eds. Spier, R.E., Griffiths, J.B. & Meignier, B. London: Butterworth-Heinemann, pp. 631–3.

Barford, J., Philips, P., Varga, E. & Harbour, C. (1992), A simulation of hybridoma growth including sugar and amino acid profiles. In *Animal cell technology: developments, processes and products*, eds. Spier, R.E., Griffiths, J.B. & MacDonald, C., London: Butterworth-Heinemann, pp. 230–2.

Barr, P.J. (1991), Mammalian subtilisins: the long sought dibasic processing endoproteases. *Cell 66*: 1–3.

Bebbington, C.R. & Hentschel, C.C.G. (1987), The use of vectors based on gene amplification for the expression of cloned genes in mammalian cells. In *DNA cloning: a practical approach*, Vol. 3, ed. Glover, D.M. Oxford: IRL Press, pp. 163–88.

Bebbington, C.R., Renner, G., Thomson, S., King, D., Abrams, D. & Yarranton, G.T. (1992), High level expression of a recombinant antibody from myeloma cells using a glutamine synthetase gene as an amplifiable selectable marker. *Bio/Technology 10*: 169–75.

Beitel, L.K., McArthur, J.G. & Stenners, C.P. (1991), Sequence requirements for the stimulation of gene amplification by a mammalian genomic element. *Gene 102*: 149–56.

Bell, S.L., Bebbington, C.R., Bushell, M.E., Sanders, P.G., Scott, M.F., Spier, R.E. & Wardell, J.N. (1991), Genetic engineering of cellular physiology. In *Production of biologicals from animal cells in culture*, eds. Spier, R.E., Griffiths, J.B. & Meignier, B. London: Butterworth-Heinemann, pp. 304–6.

Bellgardt, K.H., Meyer, H.D., Kuhlmann, W., Schugerl, K. & Thoma, M. (1984), On-line estimation of biomass and fermentation parameters by a Kalman filter during a cultivation of *S. cerevisiae*. In *3rd European Congr. Biotechnol.* Vol. 2, Munchen: Verlag Chimie, pp. 607–15.

Bendig, M.M., Stephens, P.E., Crockett, M.I. & Hentschel, C.C.G. (1987), Mouse lines that use heat shock promoters to regulate the expression of tissue plasminogen activator. *DNA 6*: 343–52.

Berg, G.J., & Bodecker, B.G.D. (1988), Employing a ceramic matrix for the immobilisation of mammalian cells in culture. In *Animal cell biotechnology*, Vol. 3, eds. Spier, R.E. & Griffiths, J.B. London: Academic Press, pp. 321–35.

Birch, J.R. & Cartwright, T. (1982), Environmental factors influencing the growth of animal cells in culture. *J. Chem. Tech. Biotechnol. 32*: 313–17.

Birch, J.R., Boraston, R. & Ward, L. (1985), Bulk production of monoclonal antibodies in fermentors. *Trends Biotechnol. 3*: 162–6.

Birch, J.R. & Boraston, R.C. (1987), *World Patent Application* WO 87/00195.

Bliem, R., Oakley, R., Taiariol, V., Matsuoka, K. & Long, J. (1991), Considerations in the design, development and scale up of glass bead packed reactors. In *Production of biologicals from animal cells in culture*, eds. Spier, R.E., Griffiths, J.B. & Meignier, B. London: Butterworth-Heinemann, pp. 407–13.

Blochlinger, K. & Diggelmann, H. (1984), Hygromycin B phosphotransferase as a selectable marker for DNA transfer experiments with higher eucaryotic cells. *Mol. Cell. Biol. 4*: 2929–31.

Blom van Assendelft, G., Hanscombe, O., Grosveld, F. & Greaves, D.R. (1989), The beta-globin dominant control region activates homologous and heterologous promoters in a tissue specific manner. *Cell 56*: 969–77.

Bloomquist, B.T., Eipper, B.A. & Manns, R.E. (1992), Prohormone converting enzymes: regulation and evaluation of function using antisense RNA. *Mol. Endo. 5*: 2014–24.

Bluml, G., Reiter, M., Zach, N., Gaida, C., Schmatz, C., Borth, N., Hohenwarter, O. & Katinger, H. (1992a), High density aggregate culture of recombinant CHO cells in fluidised bed bioreactors. In *Animal cell technology: developments, processes and products*, eds. Spier, R.E., Griffiths, J.B. & MacDonald, C. London: Butterworth-Heinemann, pp. 421–3.

Bluml, G., Reiter, M., Zach, N., Gaida, C., Schmatz, C., Strutzenberger, K., Mohr, T., Rauschert, B. & Katinger, H. (1992b), Development of a new type of macroporous carrier. In *Animal cell technology: developments, processes and products*, eds. Spier, R.E., Griffiths, J.B. & MacDonald, C. London: Butterworth-Heinemann, pp. 501–4.

Boag, A.H. & Sutton, M.V. (1987), Microencapsulation of human fibroblasts in a water insoluble polyacrylate. *Biotechnol. Bioeng. 30*: 954–62.

Bognar, E.A., Pugh, G.G. & Lydersen, B.K. (1983), Large scale production of BHK21 cells using the Opticell system. *J. Tissue Cult. Meth. 8*: 147–54.

Boraston, R., Thompson, P.W., Garland, S. & Birch, J.R. (1983), Growth and oxygen requirements of antibody producing mouse hybridoma cells in suspension culture. *Dev. Biol. Stand. 55*: 103–111.

Boschetti, E. (1991), Isolation of biologicals: leachables and cleaning in place becomes a crucial issue in liquid chromatography. In *Production of biologicals from animal cells in culture*, eds. Spier, R.E., Griffiths, J.B. & Meignier, B. London: Butterworth-Heinemann, 639–49.

Boshart, M., Weber, F., Jahn, G. *et al.* (1985), A very strong enhancer is located upstream of an immediate early gene of cytomegalovirus. *Cell 41*: 521–30.

Bostock, C.J. & Allshire, R.C. (1986), Comparison of methods for introducing vectors based on bovine papilloma virus -1 DNA into mammalian cells. *Soma. Cell. Mol. Gen. 12*: 157–366.

Bowes, K. & Bentley, C. (1993), The use of lactate dehydrogenase as an accurate method for determining viable and non-viable cell counts in a CHO population. Paper presented at ESACT UK meeting, Birmingham 7/1/93.

Brenner, H. (1966), Hydrodynamic resistance of particles. *Adv. Chem. Eng. 6*: 377–403.

Brose, D. & van Eikeren, P. (1990), A membrane-based method for removal of toxic ammonia from mammalian cell culture. *Appl. Biochem. Biotechnol. 24/25*: 457–68.

Browne, M.J., Carey, J.E., Chapman, C.G. *et al* (1988), A tissue-type plasminogen activator mutant with prolonged clearance *in vivo*. *J. Biol. Chem 263*: 1599–1602.

Builder, S.E., Van Reis, R., Paoni, N.F., Field, M. & Ogez, J.R. (1989), Process development for products derived from the new biotechnology. In *Advances in animal cell biology and technology for bioprocesses*, eds. Spier, R.E., Griffiths, J.B., Stephenne, J. & Crooy, P.J. London: Butterworth & Co., pp. 452–62.

Buntemeyer, H., Bodeker, B.G.D. & Lehmann, J. (1987), Membrane stirrer reactor for bubble free aeration and perfusion. In *Modern approaches to animal cell technology*, eds. Spier, R.E. & Griffiths, J.B. London: Butterworths pp. 411–19.

Burns, P.A., Jack, A., Neilson, F., Hadlow, S. & Balmain, A. (1991), Transformation of mouse skin endothelial cells in vivo by direct application of plasmid DNA encoding the T24 H-ras oncogene. *Oncogene 6*: 1973–8.

Butler, M., Hassell, T., Doyle, C., Gleave, S. & Jennings, P. (1991), The effect of metabolic by-products on animal cells in culture. In *Production of biologicals from animal cells in culture*, eds. Spier, R.E., Griffiths, J.B. & Meignier, B. London: Butterworth-Heinemann, pp. 226–8.

Byrn, R.A., Mordenti, J., Lucas, C., Smith, D., Marsters, S.A., Johnson, J.S., Cossum, P., Chamow, S.M., Wurm, F.M., Gregory, T., Groopman, J.E. & Capon, D.J. (1989), Biological properties of a CD4 immunoadhesin. *Nature 344*: 667–70.

Cadeiro, M.T. & Curling, E.M.A. (1992), Production and use of monoclonal antibodies of different isotypes to study the effects of glycosylation on antibody production in culture. Paper presented at ESACT UK, Manchester, 23rd April 1992.

Cambier, P., Van der Werf, F., Larsen, G.F. & Collen, D. (1988), Pharmacokinetics and pharmacological properties of a non-glycosylated mutant of human tissue type plasminogen activator lacking the finger and growth factor domains, in dogs with copper coil-induced coronary artery thrombosis. *J. Cardiovasc. Pharmacol.* 2: 468–72.

Campo, M.S. (1987), Bovine papilloma virus DNA: a eucaryotic cloning vector. I. DNA cloning: a practical approach, Vol. 2, ed. Glover, D.M. Oxford: IRL Press Oxford, pp. 213–38.

Campo, M.S. & Spandidos, D.A. (1983), Molecularly cloned bovine papilloma virus DNA transforms mouse fibroblasts *in vitro*. *J. Gen. Virol. 64*: 549–57.

Cartwright, T. (1987), Isolation and purification of products from animal cells. *Tibtech 5*: 25–30.

Cartwright, T. (1992), The production of tPA from animal cells. In *Animal Cell Biotechnology* Vol.5, eds Spier, R.E. & Griffiths, J.B. London: Academic Press, pp. 217–45.

Cartwright, T. & Crespo, A. (1991), Production of a pharmaceutical enzyme: Animal cells or E. coli. In *Production of biologicals from animal cells in culture*, eds. Spier, R.E., Griffiths, J.B. & Meignier, B. London: Butterworth-Heinemann, pp. 669–69.

Cartwright, T. & Duchesne, M. (1985), Purification of products from animal cells. In *Animal cell biotechnology*, Vol.2, eds. Spier, R.E. & Griffiths, J.B. London: Academic Press, pp. 151–84.

Cartwright, T. & Swain, D. (1974), Interferon assay by enzyme release assessment of virally induced cytopathic effect. *Interferon Scientific Memoranda* I-455/1.

Cattaneo, M.V. & Luong, J.H.T. (1993), Monitoring glutamine in animal cell cultures using a chemoluminescence fibre optic biosensor. *Biotech. Bioeng. 41*: 659–65.

Chang, E.H., Ellis, R.W., Scolnik, E.M. & Lowry, D.R. (1980), Transformation by cloned Harvey murine sarcoma virus DNA: efficiency increased by long terminal repeat DNA. *Science 210*: 1249–51.

Chenciner, N., Delpeyroux, F., Israel, N., Lambert, M., Lim, A. & Houssais, J.F. (1990), Enhancement of gene expression by somatic hybridization with primary cells: high level synthesis of the hepatitis B surface antigen in monkey Vero cells by fusion with primary hepatocytes. *Bio/Technology 8*: 858–62.

Chengalvala, M.V.R., Lubeck, M.D., Selling, B.J., Natuk, R.J., Hsu, K.H.L., Mason, B.B., Chanda, P.K., Bhat, R.A., Bhat, B.M., Mizutani, S., Davis, A.R. & Hung, P.P. (1991), Adenovirus vectors for gene expression. *Current Opinion Biotech. 2*: 718–22.

Cherry, R.S. & Aloi, L.E. (1991), Calcium transients in Sf-9 insect cells exposed to fluid stresses representative of turbulent eddies. In *Animal cell technology: developments, processes and products*, eds. Spier, R.E., Griffiths, J.B. & Meignier, B. London: Butterworth-Heinemann, pp. 206–11.

Chou, S. & Merigan, T.C. (1983), Rapid detection and quantitation of human cytomegalovirus in urine through DNA hybridisation. *N. Engl. J. Med. 308*: 921–5.

Cockett, M.I., Bebbington, C.R. & Yarranton, G.T. (1990), High level expression of tissue inhibitor of metalloproteinases in Chinese hamster ovary cells using glutamine synthetase gene amplification. *Bio/Technology 8*: 662–7.

Colbere-Gerapin, G., Horodniceau, F., Kourilsky, P. & Gerapin, A.C. (1981), A new dominant hybrid selective marker for higher eucaryotic cells. *J. Mol. Biol. 150*: 1–14.

Collee, J.G. (1991), Bovine spongiform encephalopathy. *Med. Lab. Sci. 48*: 296–302.

Cossons, N.H., Hayter, P.M., Tuite, M.F. & Jenkins, N. (1991), Stability of amplified DNA in Chinese hamster ovary cells. In *Production of biologicals from animal cells in culture*, eds. Spier, R.E., Griffiths J.B. & Meignier, B. London: Butterworth-Heinemann, pp. 309–14.

Cousins, R.B., Gergen, R. & Gerner, F.J. (1992), A dual fibre membrane reactor for the cultivation of mammalian cells. In *Animal cell technology: developments, processes and products*, eds. Spier, R.E., Griffiths, J.B. & McDonald, C. London: Butterworth-Heinemann, pp. 538–40.

Croughan, M.S., Hamel, J.F. & Wang, D.I.C. (1987), Hydrodynamic effects on animal cells in microcarrier culture. *Biotechnol. Bioeng. 29*: 130–41.

Croughan, M.S., Sayrre, E.S. & Wang, D.I.C. (1988), Viscous reduction of turbulent damage in animal cell culture. *Biotechnol. Bioeng. 33*: 862–72.

Culip, J.S., Johanssen, H., Hellring, B., Beck, J., Matthews, T.J., Delers, A. & Rosenberg, M. (1991), Regulated expression allows high level production and secretion of HIV-1 gp 120 envelope glycoprotein in *Drosophila* Schneider cells. *Bio/Technology 9*: 173–8.

Curling, E.M.A., Hayter, P.M., Baines, A.J., Bull, A.T., Gull, K., Strange, P. & Jenkins, M. (1990), Recombinant human interleukin-8: differences in glycosylation and proteolytic processing lead to heterogeneity in batch culture. *Biochem. J. 272*: 333–7.

Darling, A.J., Smith, K.T., Manson, J. & Hopes, J. (1992), Prions: validation of processes for the removal and/or inactivation of these agents. In *Animal cell technology: developments, processes and products*, eds. Spier, R.E., Griffiths, J.B. & MacDonald, C. London: Butterworth-Heinemann, pp. 671–4.

Davidson, S.K. & Hunt, L.A. (1985), Sindbis virus proteins are abnormally glycosylated in CHO cells deprived of glucose. *J. Gen. Virol. 66*: 1457–68.

Dean, R.T., Jessup, W. & Roberts, C.R. (1984), Effects of exogenous amines on mammalian cells with particular reference to membrane flow. *Biochem. J. 217*: 27–40.

Dhainhaut, F., Pouget, L., Ricter-Hers, M.J. & Mignot, G. (1991), Optimisation of human anti-rhesus IgG production. In *Production of biologicals from animal cells in culture*, eds. Spier, R.E., Griffiths, J.B. & Meignier, B. London: Butterworth-Heinemann, pp. 495–7.

Di Maio, D., Corbin, V., Sibley, E. & Maniatis, T. (1984), High level expression of a cloned HLA heavy chain gene introduced into mouse cells on a bovine papilloma virus vector. *Mol. Cell Biol. 4:* 340–50.

Di Maio, D., Treisman, R. & Maniatis, T. (1982), Bovine papilloma virus vector that propagates as a plasmid in both mouse and bacterial cells. *Proc. Natl. Acad. Sci. USA 79*: 4030–4.

Dodge, T.C., Ji, G.Y. & Hu, W.S. (1987), Loss of viability in hybridoma cell culture. A kinetic study. *Enzyme Microb. Technol. 9*: 600–7.

Dorner, A.J., Krane, M.G. & Kaufman, R.J. (1988), Reduction of endogenous GRP78 levels improves secretion of a heterologous protein in CHO cells. *Mol. Cell. Biol. 8*: 4063–70.

Dorner, A.J., Wasley, L.C. & Kaufman, R.J. (1989), Increased synthesis of secreted proteins induces expression of glucose regulated proteins in butyrate-treated Chinese hamster ovary cells. *J. Biol. Chem. 264*: 20602–7.

Dorner, A.J., Wasley, L.C. & Kaufman, R.J. (1990), Protein dissociation from GRP78 and secretion are blocked by depletion of cellular ATP levels. *Proc. Natl. Acad. Sci. USA 87*: 7429–32.

Dube, S., Fisher, J.W. & Powell, J.S. (1988), Glycosylation at specific sites of erythropoietin is essential for biosynthesis, secretion and biological functions. *J. Biol. Chem. 236*: 17516–32.

Duff, R.G. (1985), Microencapsulation technology: a novel method for monoclonal antibody production. *Trends Biotechnol. 3*: 167–70.

Durfor, C.N. & Scribner, C.L. (1991), An FDA perspective of manufacturing changes for products for human use. *Ann. N.Y. Acad. Sci. 665*: 356–63.

Eagle, H. (1959), Amino acid metabolism in mammalian cell cultures. *Science 130*: 432–7.

Earle, W.R., Bryant, J.C., & Schilling, E.L. (1954), Certain factors limiting the size of tissue culture and the development of massive cultures. *Ann. N. Y. Acad. Sci. 58*: 1000–1011.

Elbein, A.D. (1987), Inhibition of the biosynthesis of N-linked oligosaccharide chains. *Ann. Rev. Biochem. 56*: 497–534.

Esclade, L.R.J., Stephane, C. & Peringer, P. (1991), Influence of the screen material on the fouling of spin filters. *Biotechnol. Bioeng. 38*: 159–68.

Favre, E. & Thaler, T. (1992), An engineering analysis of rotating sieves for hybridoma cell retention in stirred tank bioreactors. Paper presented at Cell Culture Engineering III, Palm Court, Florida, Feb 2–7, 1992.

Fazekas de St Groth, S. (1983), Automated production of monoclonal antibodies in a cytostat. *J. Immunol. Meth. 57*:121–36.

Federspeil, N.A., Beverley, S.M., Schilling, J.W. & Schimke, R.T. (1984), Novel DNA rearrangements are associated with dihydrofolate reductase gene amplification. *J. Biol. Chem. 259*: 9127–40.

Feng, B., Shriber, S.K. & Max, S.R. (1990), Glutamine regulates glutamine-synthetase expression in skeletal muscle cells in culture. *J. Cell. Physiol. 145*: 376–80.

Fenge, C., Buzsaky, F., Fraune, E. & Lindner-Olsson, E. (1992), Evaluation of a spin filter during perfusion culture of recombinant CHO cells. In *Animal cell technology: developments, processes and products*, eds. Spier, R.E., Griffiths, J.B. & MacDonald, C. London: Butterworth-Heinemann, pp. 429–32.

Fertig, G., Kloppinger, M. & Miltenburger, H.G. (1990), Cell cycle kinetics of insect cell cultures compared to mammalian cell cultures. *Exp. Cell. Res. 189*: 208–12.

Finn, G.K., Kurz, B.W., Chang, R.Z. & Reis, R.J.S. (1989), Homologous plasmid recombination is elevated in imortally transformed cells. *Mol. Cell. Biol. 9*: 3377–84.

Fleischaker, R.J. & Sinskey, A.J. (1981), Oxygen demand and supply in cell culture. *Eur. J. Appl. Microbiol. Biotechnol. 12*: 193–7.

Frame, K.K. & Hu, W-S. (1990), The loss of antibody production in continuous culture of hybridoma cells. *Biotech. Bioeng. 35*: 469–76.

Frame, K.K. & Hu, W-S. (1991), Kinetic study of hybridoma cell growth in continuous culture, 1. A model for non-producing cells. *Biotechnol. Bioeng. 37*: 55–64.

Friedman R.L. (1985), Expression of human adenosine deaminase using a transmissable murine retrovirus vector. *Proc. Natl. Acad. Sci. USA 82*: 703–7.

Froud, S.J., Clements, G.J., Doyle, M.E., Harris, E.L.V., Lloyd, C., Murray, P., Stephens, P.E., Thompson, S. & Yarranton, G.T. (1991), Development of a process for the production of HIV 1 gp120 from recombinant cell lines. In *Production of biologicals from animal cells in culture*, eds. Spier, R.E., Griffiths, J.B. & Meignier, B. London: Butterworth-Heinemann, pp. 110–15.

Ganne, V., Guerin, P., Faure, T. & Mignot, G. (1991), Increased expression of Factor VIII by butyrate in Chinese hamster ovary cells. In *Production of biologicals from animal cells in culture*, eds. Spier, R.E., Griffiths, J.B. & Meignier, B. London: Butterworth-Heinemann, pp. 104–6.

Gardner, A.R., Gainer, J.L. & Kirwan, D.J. (1990), Effects of stirring and sparging on cultured hybridoma cells: the protective effect of Pluronic F68. *Biotechnol. Bioeng. 35*: 940–7.

Gillies, S.D., Dorai, H., Weslowski, J., Majeau, G., Young, D., Boyd, J., Gardner, J. & James, K. (1989), Expression of human anti-tetanus toxoid antibody in transfected murine myeloma cells. *Bio/Technology 7*: 799–804.

Glacken, M. W. (1988), Catabolic control of mammalian cell culture. *Bio/Technology 6*: 1041–50.

Glacken, M. W., Fleischaker, R. J. & Sinskey A.J.(1986), Reduction of waste product excretion via nutrient control: possible strategies for maximising product and cell yields on serum in culture of mammalian cells. *Biotechnol. Bioeng. 28*: 1376–89.

Glacken, M.W., Adema, E. & Sinskey, A.J. (1988), Mathematical description of hybridoma kinetics: II, The relationship between thiol chemistry and the degradation of serum activity. *Biotechnol. Bioeng. 33*: 440–5.

Goeddel, D., Kleid, D., Bolivar, F., Heyneker, H., Yarisura, D., Crea, R., Hirose, T., Kraszewski, A., Itakura, K. & Riggs, A. (1979), Expression in *E. coli* of chemically synthesised genes for human insulin. *Proc. Natl. Acad. Sci. USA 76*: 106–10.

Goergen, J.L., Martial, A., Marc, A. & Engasser, J.M. (1991), A kinetic model for the influence of serum in batch and continuous hybridoma cultures. In

Production of biologicals from animal cells in culture, eds. Spier, R.E., Griffiths, J.B. & Meignier, B. London: Butterworth-Heinemann, pp. 634–6.

Goetghebeur, S. & Hu, W-S. (1991), Cultivation of anchorage dependent animal cells in microsphere induced aggregate culture. *Appl. Microbiol. Biotechnol. 34*: 735–41.

Goldberg, A.L. & Dice, J.F. (1974), Intracellular protein degradation in mammalian and bacterial cells. *Ann. Rev. Biochem. 43*: 835–69.

Goldblum, S., Bae, Y.K., Hink, W.F. & Chalmers, J. (1990), Protective effect of methylcellulose and other polymers on insect cells subjected to laminar shear stress. *Biotechnol. Prog. 6*: 383–90.

Goochee, C.F. & Monica, T. (1990), Environmental effects on protein glycosylation. *Bio/Technology 8*: 421–7.

Gorman, C. (1986), High efficiency gene transfer into mammalian cells. In *DNA cloning*, Vol.2, ed. Glover, D.M. Oxford: IRL Press, pp. 143–65.

Gorman, C.M., Merlino, G.T., Willingham, M.C., Pastan, I. & Howard, B.H. (1992), Sarcoma virus long terminal repeat is a strong promoter when introduced into a variety of eukaryotic cells by DNA mediated transfection. *Proc. Natl. Acad. Sci. USA 83*: 6777–81.

Gospodarowicz, D., Ferrara, N., Schweigerer, L.& Neufeld, G. (1987), Structural characterization and biological properties of fibroblast growth factor. *Endocrinol. Rev. 8*: 95–114.

Goto, M., Akai, K., Murakami, A., Hasimoto, C., Tsuda, E., Ueda, M., Kawanishi, G., Takahashi, N., Ishimoto, A., Chiba, H. & Saski, R. (1988), Production of recombinant erythropoeitin in mammalian cells-host cell dependency of the biological activity of the cloned glycoprotein. *Bio/Technology 6*: 67–71.

Graham, F.L. (1990), Adenovirus as expression vectors and recombinant vaccines. *Trends Biotech. 8*: 85–7.

Greaves, D.R., Wilson, F.D., Lang, G. & Kioussis, D. (1989), Human CD2 3' flanking sequence confers high level T cell specific, position independent gene expression in transgenic mice. *Cell 56*: 979–86.

Green, L., Schlaffer, I. & Wright, K., (1986), Cell cycle-dependent expression of a stable episomal human histone gene in a mouse cell. *Proc. Natl. Acad. Sci. USA 83*: 2315–19.

Gribben, J.G., Devereux, S., Thomas, N.B.S., Keim, M., Jones, H.M., Goldstone, A.H. & Linch, D.C. (1990), Development of antibodies to unprotected glycosylation sites in recombinant human GM-CSF. *Lancet 335*: 434–437.

Griffiths, B. (1984), The use of oxidation-reduction potential (ORP) to monitor growth during a cell culture. *Develop. Biol. Standard. 55*: 113–16.

Grosveld, F., Blom van Assendelft, G., Greaves, D.R. & Kollias, G. (1989), Position-independent high level expression of the human beta-globin gene in transgenic mice. *Cell 51*: 975–83.

Grummt, F., Paul, D. & Grummt, I. (1977), Regulation of ATP pools, rRNA and DNA synthesis in 3T3 cells in reponse to serum or hypoxanthine. *Eur. J. Biochem. 76*: 7–12.

Grumpp, G.E. & Stephanopoulos, G. (1991), Development and scale up of controlled secretion processes for improved product recovery in animal cell culture. *Ann. N.Y. Acad. Sci. 665*: 81–93.

Guarino, L.A. (1989), Enhancers of early gene expression. In *Invertebrate cell system applications,* Vol. 1, ed. J Mitsuhashi, Boca Raton, Florida: CRC Press, pp. 211–19.
Gupta, R.K. (1987), ^{23}Na NMR spectroscopy of intact cells and tissues. In *NMR spectroscopy of cells and organisms*, Vol. 2, ed. Gupta, R.K. Boca Raton, Florida: CRC Press, pp. 1–32.
Haber, D.A. & Schimke, R.T. (1982), Chromosome mediated transfer, and amplification of an altered dihydrofolate reductase gene. *Somatic. Cell Genet. 8*: 499–508.
Hamamoto, K., Ishimaru, K. & Tokashiki, M. (1989), Perfusion culture of hybridoma cells using a centrifuge to separate cells from culture mixture. *J. Ferment. Bioeng. 67*: 190–4.
Hamamoto, K., Tokashiki, Y., Ichikawa, Y. & Murakami, Y. (1987), High cell density culture of a hybridoma using perfluorocarbon to supply oxygen. *Agric. Biol. Chem. 5*: 3415–16.
Hambach, B., Biselli, M., Runstandler, P.W. & Wandrey, C. (1992), Development of a reactor-integrated aeration sytem for cultivation of animal cells in fluidised beds. In *Animal cell technology: developments, processes and products*, eds. Spier, R.E., Griffiths, J.B. & MacDonald, C. London: Butterworth-Heinemann, pp. 381–4.
Hamer, D.H. & Walling, M.J. (1982), Regulation *in vivo* of a cloned mammalian gene: cadmium induces the transcription of a mouse metallothionein gene in SV40 vectors. *J. Molec. Appl. Genet. 1*: 273–88.
Handa-Corrigan, A., Kwasowski, P., Zhu, J.Y., Zhang, S., Zhang, L., Duggel, N., Odeby, T. & Dagwell, R. (1992), Pluronic absorption effects in cell culture medium. In *Animal cell technology: developments, processes and products*, eds. Spier, R.E., Griffiths, J.B. & MacDonald, C. London: Butterworth-Heinemann, pp. 117–21.
Harbour, C., Spencer, J., Woodhouse, G. & Barford, J.P. (1992), The development of appropriate virus models for validating therapeutic protein purification processes. In *Animal cell technology: developments, processes and products*, eds. Spier, R.E., Griffiths, J.B. & MacDonald, C. London: Butterworth-Heinemann, pp. 664–67.
Harris, S.I., Balban, R.S., Barnett, L. & Mandel, L.J. (1981), Mitochondrial respiratory capacity and Na^+-dependent and K^+-dependent adenosine triose phosphate-mediated ion transport in the intact renal cell. *J. Biol. Chem. 256*: 319–28.
Hartman, S.C. & Mulligan, R.C. (1988), Two dominant-acting selectable markers for gene transfer studies in mammalian cells. *Proc. Natl. Acad. Sci. USA 85*: 8047–51.
Hassel, T., Brand, H., Renner, G., Westlake, A. & Field, R.P. (1992), Stability of production of recombinant antibodies from glutamine synthetase amplified CHO and NSO cell lines. In *Animal cell technology: development processes and products*, eds. Spier, R.E., Griffiths, J.B. & MacDonald, C. London: Butterworth-Heinemann, pp. 42–7.
Hatcher, V.B., Wertheim, M.S., Rhee, C.Y., Tsien, G. & Burk, P.G. (1977), Relationship between cell surface protease activity and doubling time in various normal and transformed cells. *Biochem. Biophys. Res. Comm. 76*: 602–8.

Hecht, V., Bischoff, L. & Gerth, K. (1990), Hollow fibre supported gas membrane for *in situ* removal of ammonium during an antibody fermentation. *Biotechnol. Bioeng. 35*: 1042–50.

Hendricks, M.B., Banker, M.J. & McLaughlan, M. (1988), A high efficiency vector for expression of foreign genes in myeloma cells. *Gene 64*: 43–51.

Hendricks, M.B., Luchette, C.A. & Banker, M.J. (1989), Enhanced expression of an immunoglobulin-based vector in myeloma cells mediated by co-amplification with a mutant dihydrofolate reductase gene. *Bio/Technology 7*: 1271–4.

Hill-Perkins, M.S. & Possee, R.D. (1990), A baculovirus expression vector derived from the basic protein promoter of *Autograph californica* nuclear polyhedrosis virus. *J. Gen. Virol. 71*: 971–6.

Himmelfarb, P., Thayer, P.S. & Martin, H.E. (1969), Spin filter culture: the propagation of animal cells in suspension. *Science 164*: 555–7.

Hippenmeyer, P. & Highkin, M. (1993), High level, stable production of recombinant proteins in mammalian cell cultures using the herpes virus VP16 transactivator. *Bio/Technology 11*: 1037–41.

Hodgson, J. (1991), Checking sources: the serum supply secret. *Bio/Technology 9*: 1320–4.

Hodgson, J. (1993), Foetal bovine serum revisited. *Bio/Technology 11*: 49–53.

Holmberg, A., Ohlson, S. & Lundgren, T. (1991), Rapid monitoring of monoclonal antibodies in cell culture medium by HPLC. In *Production of biologicals from animal cells in culture*, eds. Spier, R.E., Griffiths, J.B. & Meignier, B. London: Butterworth-Heinemann, pp. 594–6.

Horaud, F. (1991), Biologicals and regulatory aspects in the EEC. In *Production of biologicals from animal cells in culture*, eds. Spier, R.E., Griffiths, J.B. & Meignier, B. London: Butterworth-Heinemann, pp. 759–66.

Hsieh, P., Rosner, M.R. & Rabbin, P.W. (1983), Host-dependent variation of asparagine-linked oligosaccharides at individual glycosylation sites of sindbis virus glycoproteins. *J. Biol. Chem. 258*: 2548–54.

Hsu, Y-L. & Chu, I-M. (1992), Poly(ethyleneimine)-reinforced liquid core capsules for the cultivation of hybridoma cells. *Biotechnol. Bioeng. 40*: 1300–8.

Hu, W-S., Scholz, M.T., Favre, E. & Seamany, T.C. (1991), A new look at animal cell bioreactor development. In *Production of biologicals from animal cells in culture*, eds. Spier, R.E., Griffiths, J.B. & Meignier, B. London: Butterworth-Heinemann, pp. 370–8.

Hwang, C. & Sinskey, A.J. (1991), The role of oxidation-reduction potential in monitoring the growth of cultured cells. In *Production of biologicals from animal cells in culture*, eds. Spier, R.E., Griffiths, J.B. & Meignier, B. London: Butterworth-Heinemann, pp. 548–67.

Hwang, L-H.S. & Gilbou, E. (1984), Expression of genes introduced into cells by retroviral infection is more efficient than that of genes introduced by DNA transfection. *J. Virol. 50*: 417–24.

Iatrou, K., Meidinger, R.G. & Goldsmith, M.R. (1989), Recombinant baculoviruses as vectors for identifying protein encoded by intron containing members of complex multi-gene families. *Proc. Natl. Acad. Sci. USA 86*: 9129–33.

Iio, M. & Moriyama-Takashima, A. (1985), Effect of ammonia, a toxic metabolite on cultured cells and some trials removing it from culture. *Kumamoto Joshi Daigaku Gajujutsu Kiyo 37*: 118–27.

Imamura, T., Crespi, C. L., Thilly, W.G. & Brunengraber, H. (1982), Fructose as a carbohydrate source yields stable pH and redox parameters in microcarrier cultures. *Anal. Biochem. 124*: 353–8.

Jäger, V. (1991), A novel perfusion system for the large scale cultivation of animal cells based on a continuous flow centrifuge. In *Animal cell technology: developments, processes and products*, eds. Spier, R.E., Griffiths, J.B. & Meignier, B. London: Butterworth-Heinemann, pp. 397–401.

Jähsen, J., Martens, D. & Tramper, J. (1991), Lethal events during gas sparging in animal cell culture. *Biotechnol. Bioeng. 37*: 484–90.

Jalanko, A., Pirhonen, J., Pohl, G. & Hansson, L. (1990), Production of human tissue-type plasminogen activator in different mammalian cell lines using an Epstein-Barr virus vector. *J. Biotechnol. 15*: 155–68.

Jan, D.C.H., Emery, A.N. & Al-Rubeai, M. (1992), Optimisation of spin filter performance in the intensive culture of suspended cells. In *Animal cell technology: developments, processes and products*, eds. Spier, R.E., Griffiths, J.B. & MacDonald, C. London: Butterworth-Heinemann, pp. 448–51.

Jarvis, D.L., Fleming, J-O. G.W., Kovacs, G.R., Summers, M.D. & Guarino, L.A. (1990), Use of early baculovirus promoters for continuous expression and efficient processing of foreign gene products in stably transformed lepidopteran cells. *Bio/Technology 8*: 950–5.

Jayme, D.W. & Greenwold, D.J. (1991), Media selection and design. *Bio/ Technology 9*: 716–21.

Jeffreys, A.J., Wilson, V. & Thein, S.L. (1985), Hypervariable "mini-satellite" regions in human DNA. *Nature 316*: 67–73.

Jervis, E., Lee, D.W. & Kilburn, D.G. (1991), Application of the PCFIA to rapid off-line process monitoring. In *Production of biologicals from animal cells in culture*, eds. Spier, R.E., Griffiths, J.B. & Meignier, B. London: Butterworth-Heinemann, pp. 597–9.

Jobses, J., Martens, D. & Tramper, J. (1991), Lethal events in gas sparging in animal cell culture. *Biotechnol. Bioeng. 37*: 484–90.

Johansson, B.L., Hellberg, U. & Wennburg, O. (1987), Determination of the leakage from phenyl-Sepharose CL-4B, phenyl-Sepharose FF and phenyl-Superose in bulk and column experiments. *J. Chromat. 403*: 85–98.

Johnston, M.D. (1980), Enhanced production of interferon from human lymphoblastoid cells pre-treated with sodium butyrate. *J. Gen. Virol. 50*: 191–9.

Jordan, M., Renner, W., Sucker, H., Leist, C. & Eppenberger, H.M. (1992), Tuning of shear sensitivity of CHO cells and its correlation with the size distribution of cell aggregates. In *Animal cell technology: developments, processes and products*, eds. Spier, R.E., Griffiths, J.B. & MacDonald, C. London: Butterworth-Heinemann, pp. 418–20.

Kagawa, Y., Takasiki, S., Utsami, J., Hosoi, K., Shimizu, H., Kochibe, N. & Kobata, A. (1988), Comparative study of the asparagine-linked sugar chains of natural interferon beta 1 and recombinant interferon beta 1 produced in three different mammalian cells. *J. Biol. Chem. 263*: 17508–15.

Kane, J. & Hartley, D. (1988), Formation of recombinant protein inclusion bodies in *E. coli. Tibtech 6*: 95–101.

Kassenbrock, C.K., Garcia, P.D., Walter, P., & Kelly, R.B. (1988), Heavy-chain binding protein recognises aberrant polypeptides translated *in vitro*. *Nature 333*: 90–93.

Katinger, H.W.D., Schierer, W. & Kromer, E. (1979), Massenproduktion monoklonuler Antikorper *Ger. Chem. Eng. 2*: 31–8.

Kaufman, R.J., Murtha, P., Ingolia, D.E., Young, C.Y. & Kellems, R.E. (1986), Selection and amplification of heterologous genes encoding adenosine deaminase in mammalian cells. *Proc. Natl. Acad. Sci. USA 83*: 3136–40.

Kaufman, R.J., Wasley, L.C., Davies, M.V., Wise, R.J., Israel, D.I. & Dorner, A.J. (1989), Effect of von Willebrand Factor co-expression on the synthesis and secretion of Factor VIII in Chinese hamster ovary cells. *Mol. Cell. Biol. 9*: 1233–42.

Kelly, R.B., (1985), Pathways of protein secretion in eukaryotes. *Science 230*: 25–32.

Khatter, N., Matson, R.S. & Ngo, T. (1991), Utilisation of a synthetic affinity ligand for immunoglobulin purification. *American Lab.*, May 1991.

Kilburn, D.G. & Webb, F.C. (1968), The cultivation of animal cells at controlled dissolved oxygen partial pressure. *Biotechnol. Bioeng. 101*: 801–14.

Köhler, G. & Milstein, C. (1975), Continuous cultures of fused cells secreting antibody of predefined specificity. *Nature 256*: 495–7.

Konopitzky, K., Kenzler, O. & Windhab, K. (1991), Monoclonal antibody production using an airlift fermentor system consisting of a continuous seed fermentor and a fed batch production fermentor. In *Production of biologicals from animal cells in culture*, eds. Spier, R.E., Griffiths, J.B. & Meignier, B. London: Butterworth-Heinemann, pp. 390–3.

Kornfeld, R. & Kornfeld, S. (1985), Assembly of asparagine-linked oligosaccharides. *Ann. Rev. Biochem. 54*: 631–64.

Krazek, R.A., Guillino, P.M., Kohler, P. & Dedrick, R.L. (1972), Cell culture on artificial capillaries: an approach to tissue growth in vitro. *Science 178*: 65–66.

Kruh, J. (1982), Effects of sodium butyrate, a new pharmacological agent, on cells in culture. *Mol. Cell. Biochem. 42*: 65–82.

Kunas, K.T. & Papoutsakis, E.T. (1990a), Damage mechanisms of suspended animal cells in agitated bioreactors with and without bubble entrainment. *Biotechnol. Bioeng. 36*: 466–73.

Kunas, K.T. & Papoutsakis, E.T. (1990b), The protective effect of serum against hydrodynamic damage of hybridoma cells in agitated and surface aerated bioreactors. *J. Biotech. 15*: 57–70.

Kuroda, K., Geyer, H., Geyer, R., Doerfler, W. & Klenk, H-D. (1990), The oligosaccharides of influenza virus haemagglutinin expressed in insect cells by a baculovirus vector. *Virology 174*: 418–29.

Kuroda, K., Veit, M. & Klenk, H-D. (1991), Retarded processing of influenza virus haemagglutinin in insect cells. *Virology 180*: 159–65.

Kushner, P.J., Levinson, B.B. & Goodman, H.M. (1982), A plasmid that replicates in both mouse and *E. coli* cells. *J. Mol. Appl. Genet. 1*: 527–38.

Lasch, J. & Janowski, F. (1988), Linkage stability of ligand-support conjugates under operational conditions. *Enz. Microb. Technol. 10*: 312–14.

Law, M-F., Byrne, J.C. & Howley, P.M. (1983), A stable bovine papilloma virus hybrid plasmid that expresses a dominant selective trait. *Mol. Cell. Biol. 3*: 2110–15.

Lee, D.W., Grace, J.R., Allardyce, P. & Kilburn, D.G. (1991), High intensity growth of adherent cells on a porous ceramic matrix. In *Production of biologicals from animal cells in culture*, eds. Spier, R.E., Griffiths, J.B. & Meignier, B. London: Butterworth-Heinemann, pp. 400–5.

Lee, D.W., Piret, J.M., Gregory, D., Haddow, D.J. & Kilburn, D.G. (1992), Polystyrene macroporous bead support for mammalian cell culture. *Ann. N.Y. Acad. Sci. 665*: 137–45.

Lee, E.U., Roth, J. & Paulson, J.C. (1988), Alteration of terminal glycosylation sequence on N-linked oligosaccharides of Chinese hamster ovary cells by expression of beta-galactosidase-alpha 2,6-sialyl transferase. *J. Biol. Chem. 261*: 13818 55.

Lees, G. & Onions, D. (1991), Validation of downstream processing. In *Production of biologicals from animal cells in culture*, eds. Spier, R.E., Griffiths, J.B. & Meignier, B. London: Butterworth-Heinemann, pp. 783–8.

Lehmann, J., Vorlop, J. & Buntemeyer, H. (1988), Bubble-free reactors and their development for continuous culture with cell recycle. In *Animal cell biotechnology*, eds. Spier, R.E. & Griffiths, J.B. London: Academic Press, pp. 221–37.

Leiter, K.J., Wenner, C.E. & Tomei, L.D. (1985), Correlation of quabain-sensitive ion movements with cell cycle activation. *Proc. Natl. Acad. Sci. USA 72*: 1599–1603.

Letvin, N.L., Lord, C.I., King, N.W., Wyand, M.S., Myrick, K.V. & Haseltine, W.A. (1991), Risks of handling HIV. *Nature 349*: 573.

Liptrot, C. & Gull, K. (1992), Detection of viruses in recombinant cells by electron microscopy. In *Animal cell technology: developments, processes and products*, eds. Spier, R.E., Griffiths, J.B. & MacDonald, C. London: Butterworth-Heinemann, pp. 653–6.

Litwin, J. (1991), The growth of CHO and BHK cells as suspended aggregates in serum-free medium. In *Production of biologicals from animal cells in culture*, eds. Spier, R.E., Griffiths, J.B. & Meignier, B. London: Butterworth-Heinemann, pp. 429–33.

Ljunggren, J. & Haggstrom, L. (1990), Glutamine limited fed batch culture reduced ammonium ion production in animal cells. *Biotechnol. Lett. 12*: 705–10.

Looby, D. & Griffiths, J.B. (1990), Immobilisation of animal cells in porous carrier culture. *Trends Biotechnol. 8*: 204–9.

Looney, J.E. & Hamlin, J.L. (1987), Isolation of the amplified dihydrofolate reductase domain from methotrexate-resistant Chinese hamster ovary cells. *Mol. Cell. Biol. 7*: 569–77.

Lubiniecki, A., Arathoon, R., Polastri, G., Thomas, J., Weibe, M., Garnick, R., Jones, A., Van Reis, A. & Builder, S. (1989), Selected strategies for manufacture and control of recombinant tissue plasminogen activator prepared from cell cultures. In *Advances in animal cell biology and technology for bioprocesses*, eds. Spier, R.E., Griffiths, J.B., Stephenne, J. & Crooy, P.J. London: Butterworth & Co., pp 442–51.

Lucas, C., Nelson, C., Peterson, M.L., Fries, S., Vetterlein, D., Gregory, T. & Chen, A.B. (1988), Enzyme linked immunoabsorbance assay (ELISA) for the determination of contaminants resulting from immunopurification of recombinant proteins. *J. Immunol. Meth. 113*: 113–22.

Luckow, V.A. & Summers, M.D. (1988), Trends in the development of baculovirus expression vectors. *Bio/Technology 6*: 47–55.

Lucore, C.L., Fryg, E.T.A., Nachowiak, D.A. & Sobel, B.E. (1988), Biochemical determinants of clearance of tissue type plasminogen activator from the circulation. *Circulation 77*: 906–14.

Lusky, M., Berg, L., Weihr, H. & Botchan, M. (1983), Bovine papilloma virus contains an activator of gene expression at the distal end of the early transcription unit. *Mol. Cell. Biol. 3*: 1108–22.

Madisen, L., Travis, B., Hu, S-L., & Purchio, A.F. (1987), Expression of the human immunodeficiency virus gag gene in insect cells. *Virology 158*: 248–50.

Mailly, E., Fonteix, C., Engasser, J.M. & Marc, A. (1991), Mathematical estimator for the evaluation of cell density and medium composition in hybridoma cultures. In *Production of biologicals from animal cells in culture*, eds. Spier, R.E., Griffiths, J.B. & Meignier, B. London: Butterworth-Heinemann, pp. 603–5.

Mancusco, A., Fernandez, E.J., Blanch, H.W. & Clark, D.S. (1990), A nuclear magnetic resonance technique for determining hybridoma concentration in hollow fibre bioreactors. *Bio/Technology 8*: 1282–85.

Mantei, N., Boll, W. & Weissman, C. (1979), Rabbit beta-globin mRNA production in mouse L cells transformed with a cloned rabbit beta-globin chromosomal DNA. *Nature 281*: 40–56.

Marc, A., Wagner, A., Martial, A., Goergen, J.L., Engasser, J.M., Geaugey, Y. & Pinton, H. (1991), Potential and pitfalls of using LDH release for the evaluation of animal cell death kinetics. In *Production of biologicals from animal cells in culture*, eds. Spier, R.E., Griffiths, J.B. & Meignier, B. London: Butterworth-Heinemann, pp. 569–74.

Martinelle, K. & Haggsrom, L. (1992), Potassium ions affect the ammonia/ammonium ion toxicity in animal cell cultivation. In *Animal cell technology: developments, processes and products*, eds. Spier, R.E., Griffiths, J.B. & MacDonald, C. London: Butterworth-Heinemann, pp. 163–5.

Matsushita, T., Hidaka, H., Kamashita, K., Kawakubo, Y. & Funatsuki, K. (1991), High density culture of anchorage-dependent animal cells by polyurethane foam packed bed culture systems. *Appl. Microbiol. Biotechnol. 35*: 159–64.

Matsuura, Y., Possee, R.D., Overton, H.A. & Bishop, D.H.L (1987), Baculovirus expression vectors: the requirements for high level expression of proteins, including glycoproteins. *J. Gen. Virol. 68:* 1233–50.

Matthias, P.D., Bernard, H.V., Scott, A. et al (1983), A bovine papilloma virus vector with a dominant resistance marker replicates extrachromosomally in mouse and *E. coli* cells. *EMBO J. 2*: 1487–92.

Matthiasson, E. (1983), The role of macromolecular adsorption in the fouling of ultrafiltration membranes. *J. Membrane Sci. 16*: 23–32.

McArthur, J.G. & Stanners, C.P. (1991), A genetic element that increases the frequency of gene amplification. *J. Biol. Chem. 266*: 6000–5.

McKnabb, S., Rupp, R. & Tedesco, J.L. (1989), Measuring contaminating DNA in bioreactor derived monoclonals. *Bio/Technology 7*: 343–7.

McLean, J., Hollis, M., Needham, M., Gooding, C., Hudson, K., Scanlon, D., McLaughlin, F., Grosveld, C. & Antoniou, M. (1992), A novel mammalian expression system based on the globin locus control region. In *Animal cell*

technology: developments, processes and products, eds. Spier, R.E., Griffiths, J.B. & MacDonald, C., London: Butterworth-Heinemann, pp. 68–70.
McQueen, A. & Bailey, J.E. (1990), Effect of ammonium ion and intracellular pH on hybridoma cell metabolism and antibody production. *Biotechnol. Bioeng. 35*: 1067–77.
McQueen, A., Meilhoc, E. & Bailey, J.E. (1987), Flow effects on the viability and lysis of suspended mammalian cells. *Biotechnol. Lett. 9*: 832–6.
Meneguzzi, G., Binetruy, B., Grisoni, M. & Cuzin, F. (1984), Plasmidial maintenance in rodent fibroblasts of a BPV1-pBR322 shuttle vector without immediately apparent oncogenic transformation of the recipient cells. *EMBO J. 3*: 365–71.
Merten, O-W., Keller, H., Cabanie, L., van Kan Martin, C. & Moeurs, D. (1992), Release of cellular enzymes for evaluating the dead cell number in bioreactor cultures. In *Animal cell technology: developments, processes and products*, eds. Spier, R.E., Griffiths, J.B. & MacDonald, C. London: Butterworth-Heinemann, pp. 319–24.
Mignot, G., Ganne, V., Faure, T., Pavirani, A. & van der Pol, H. (1989), The use of a statistical approach for the optimisation of culture conditions of genetically engineered cell lines. In *Advances in animal cell biology and technology for bioprocessors*, eds. Spier, R.E., Griffiths, J.B., Stephanne, J. & Crooy, P.J. London: Butterworth & Co., pp. 52–58.
Mikkelsen, J., Thomsen, J. & Ezban, M. (1991), Heterogeneity in the tyrosine sulphation of Chinese hamster ovary cell produced recombinant Factor VIII. *Biochemistry 300*: 1533–7.
Miller, A.D., Law, M-F. & Verme, I.M. (1985), Generation of helper-free amphotropic retrovirus that transduces a dominant-acting, methotrexate-resistant dihydrofolate reductase gene. *Mol. Cell. Biol. 5*: 431–7.
Miller, W.M., Wilke, C.R. & Blanch, H.W. (1987), Effects of dissolved oxygen concentration on hybridoma growth and metabolism in continuous culture. *J. Cell. Physiol 132*: 524–30.
Miller, W.M., Wilke, C. & Blanch, H.W. (1988), Transient responses of hy bridoma cells to lactate and ammonia pulse and step changes in continuous culture. *Bioprocess. Eng. 33*: 477–86.
Mitrani-Rosenbaum, S., Maroteaux, L., Mory, Y., Revel, M. & Howley, P.M. (1983), Inducible expression of the human interferon beta 1 gene linked to a bovine papilloma virus DNA vector and maintained extrachromosomally in mouse cells. *Mol. Cell. Biol. 3*: 233–40.
Miyamoto, C., Smith, G.E., Farrell-Towt, J., Summers, M.D. & Ju, G. (1985), Production of human *c-myc* protein in insect cells infected with a baculovirus expression vector. *Mol. Cell. Biol. 5*: 2860–5.
Mizrahi, A. (1975), Pluronic polyols in human lymphocyte cell line cultures. *J. Clin. Microbiol. 2*: 11–13.
Mohda, K., Whiteside, J.P. & Spier, R.E. (1992), Dissociation of MAB production from cell division using DNA synthesis inhibitors. In *Animal cell technology: developments, processes and products*, eds. Spier, R.E., Griffiths, J.B. & MacDonald, C. London: Butterworth-Heinemann, pp. 81–7.
Monegier, B., Clerc, F.F., van Dorsselaer, A., Vuilhorgne, M., Green, B. & Cartwright, T. (1990), Using mass spectrometry to characterise recombinant proteins. *Pharmac. Technol. Int. 2*: 19–30.
Moses, H.L., Coffrey, R.J., Leof, E.B., Lyons, R.M. & Keksi-Oja, J. (1987),

Transforming growth factor beta regulation of cell proliferation. *J. Cell. Physiol. Suppl. 5*: 1–7.

Moss, B. & Flexner, C. (1987), Vaccinia virus expression vectors. *Ann. Rev. Immunol. 5*: 305–24.

Mulligan, R.C. & Berg, P. (1981), Selection for animal cells that express the *E.coli* gene encoding for xanthine-guanine phosphoribosyltransferase. *Proc. Natl. Acad. Sci. USA. 78*: 2072–6.

Murhammer, D.W. & Goochee, C.F. (1990), Sparged animal cell bioreactors: mechanism of cell damage and Pluronic F-68 protection. *Biotechnol.Prog. 6*: 391–7.

Murray, K., Dickson, A.J. & Gull, K. (1992), Metabolic management of a hybridoma cell line. In *Animal cell technology: developments, processes and products*, eds. Spier, R.E., Griffiths, J.B. & MacDonald, C. London: Butterworth-Heinemann, pp. 261–3.

Myoken D. (1989), An alternative method for the isolation of NS-1 hybridomous using cholesterol auxotrophy of NS-1 mouse myeloma cells. *In Vitro 25*: 477–80.

Nakano, E.T., Cianpi, N.A. & Young, D.V. (1982), The identification of a serum viability factor for SV3T3 cells as biotin and its possible relationship to Krebs cycle activity. *Arch. Biochem. Biophys. 215*: 556–63.

Nakatani, T., Nomura, I., Horigami, K., Okamoto, M. & Yoshima, T. (1989), Functional expression of human monoclonal antibody genes directed against pseudomonal endotoxin A in mouse myeloma cells. *Bio/Technology 7*: 805–10.

Nakazawa, K., Furukawa, K., Kobata, A. & Narimitsu, H. (1991), Characterisation of a murine B1-4 galactosyl transferase expressed in COS-1 cells. *Eur. J. Biochem. 196*: 363–8.

Neilsen, L.K., Niloperbow, W., Reid, S. & Greenfield, P.F. (1991), Modelling growth of and antibody production by hybridomas in glutamine limited suspension culture. In *Production of biologicals from animal cells in culture*, eds. Spier, R.E., Griffiths, J.B. & Meignier, B. London: Butterworth-Heinemann, pp. 625–30.

Neuberger, M.S. (1983), Expression and regulation of immunoglobulin heavy chain genes transfected into lymphoid cells. *Embo J. 2*: 1373–8.

Nevaril, C. G., Lynch, E. C., Alfrey, C. P. & Hellums, J.D. (1968), Erythrocyte damage and destruction by shearing stress. *J. Lab. Clin. Med. 71*: 784–90.

Nikolai, T.J. & Hu, W-S. (1992), Cultivation of mammalian cells on macroporous microcarriers. *Enz. Microbiol. Technol. 14*: 203–8.

Nyberg, S.L., Shatford, R.A., Peshwa, M.V., White, J.G., Cerra, F.B. & Hu, W-S. (1993), Evaluation of a hepatocyte-entrapment hollow fibre bioreactor: a potential bioartificial liver. *Biotechnol. Bioeng. 41*: 194–203.

Oda, K., Ogata, S., Koriyama, Y., Yamada, E., Mifune, K. & Ikehara, Y. (1988), Tris inhibits both proteolytic and oligosaccharide processing occuring in the Golgi complex in primary cultured rat hepatocytes. *J. Biol. Chem. 263*: 12576–83.

Oka, M., Johansen, H., Okita, B., Wilson, B., Carr, S., Roberts, G., Chen, T., Strickler, J. & Mai, S. (1989), Purification and characterisation of human tPA produced in Drosophila cells. In *Advances in animal cell biology and tèchnology for bioprocesses*, eds. Spier, R.E., Griffiths, J.B., Stephanne, J. & Crooy, P.J. London: Butterworths, pp. 465–72.

Okamoto, M., Nakayama, C., Nakai, M. & Yenagi, H. (1990), Amplification and high level expression of a c DNA for human granulocyte-macrophage colony stimulating factor in human lymphoblastoid Namalwa cells. *Bio Technology 8*: 550–3.

Onions, D. & Lees, G. (1991), Evaluating the safety of murine hybridomas-new problems and new techniques. In *Production of biologicals from animal cells in culture*, eds. Spier, R.E., Griffiths, J.B. & Meignier, B. London: Butterworth-Heinemann, pp. 34–8.

Ostrowski, M.C., Richard-Foy, H., Walford, R.G., Berard, D.S. & Hager, G.L. (1983), Glucocorticoid regulation of transcription at an amplified episomal promoter. *Mol. Cell. Biol. 3*: 2045–57.

Oyaas, K., Berg, T.M., Bakke, O. & Levine, D.W. (1989), Hybridoma growth and antibody production under conditions of hyperosmotic stress. In *Advances in animal cell biology and technology for bioprocesses*, eds. Spier, R.E., Griffiths, J.B., Stephenne, J. & Crooy, P.J. London: Butterworth & Co., pp. 212–18.

Page, M.J. & Sydenham, M.A. (1991), High level expression of the humanised monoclonal antibody Campath-1H in Chinese hamster ovary cells. *Bio/Technology 9*: 64–8.

Panina, G.F. (1985), Monolayer growth systems: multiple processes. In *Animal cell biotechnology,* Vol. I, eds. Spier, R.E. & Griffiths, J.B. London: Academic Press, pp. 211–237.

Parekh, R.B., Dwek, R.A., Edge, C.J. & Rademacher, T.W. (1989), N-glycosylation and the production of recombinant glycoproteins. *Tibtech 7*: 117–21.

Park, S. & Stephanopoulos, G. (1993), Packed bed reactor with porous ceramic beads for animal cell culture. *Biotechnol. Bioeng. 41*: 25–34.

Pavlakis, G.N. & Hamer, D.H. (1983), Regulation of a metallothionein-growth hormone hybrid gene in bovine papilloma virus. *Proc. Natl. Acad. Sci. USA 80*: 397–401.

Pendse, G.J., Karkare, S. & Bailey, J.E. (1992), Effect of cloned gene dosage on cell growth and hepatitis B surface antigen synthesis and secretion in recombinant CHO cells. *Biotech. Bioeng. 40*: 119–29.

Pennock, G.O., Shoemaker, C. & Miller, L.K. (1984), Strong and regulated expression of *E.coli* beta-galactosidase in insect cells with a baculovirus vector. *Mol. Cell. Biol. 4*: 399–406.

Per, S.L., Averson, C.R. & Sito, A.F. (1989), Quantitation of residual mouse DNA in monoclonal antibody-based products. *Develop. Biol. Stand. 71*: 173–80.

Peshwa, M.V., Kyung, Y-S., McClure, D.B. & Hu, W-S. (1993), Cultivation of mammalian cells as aggregates in bioreactors: effect of calcium concentration on spatial distribution of viability. *Biotechnol. Bioeng. 44*: 179–87.

Petersen, J.F., McIntire, L.V. & Papoutsakis, E.T. (1988), Shear sensitivity of cultured hybridoma cells depends on mode of growth, culture age and metabolite concentration. *J. Biotechnol. 7*: 229–34.

Petricianni, J.C. (1985), Regulatory considerations for products derived from the new biotechnology. *Pharm. Man. 5*: 31–4.

Petricianni, J.C. (1988), Changing attitudes and actions governing the use of continuous cell lines for the production of biologicals. In *Animal cell bio-*

technology, Vol. 3, eds. Spier, R.E. & Griffiths, J.B. London: Academic Press, pp. 14–25.

Piret, J.M. & Cooney, C.L. (1990), Mammalian cell and protein distribution in ultrafiltration hollow fibre bioreactors. *Biotechnol. Bioeng. 36*: 902–10.

Piret, J.M. & Cooney, C.L. (1991), Model of oxygen transport limitations in hollow fibre bioreactors. *Biotechnol. Bioeng. 37*: 80–92.

Plackett, R.L. & Burman, J.P. (1946), The design of optimal multifactorial experiments. *Biometrica 33*: 305–25.

Podhajska, A.J., Hasan, N. & Szybalski, W. (1985), Control of cloned gene expression by promoter inversion *in vivo*. *Gene 40*: 163–8.

Preibisch, G., Isihara, H., Tripier, D. & Leineweber, M. (1988), Unexpected translation initiation within the coding region of eukaryotic genes expressed in *E. coli*. *Gene 72*: 179–86.

Pulladino, M.A., Levinson, A.B., Svedersky, P. & Oberjeski, J.F. (1987), Safety issues related to the use of recombinant DNA derived cell culture products: I Cellular components. *Develop. Biol. Standard. 66*: 13–22.

Pullen, K.F., Johnson, M.D., Philips, A.W., Ball, G.D. & Finter, N.B. (1985), Very large scale suspension cultures of mammalian cells. *Develop. Biol. Standard. 60*: 175–7.

Radlett, P.J., Pay, T.W.F. & Garland, A.J.M. (1985), The use of BHK suspension cells for the commercial production of Foot and Mouth Disease vaccine over a twenty year period. *Develop. Biol. Standard. 60*: 163–70.

Raper, J., Douglas, Y., Gordon-Walker, N. & Caulcott, C.A. (1992), Long term stability of expression of humanised monoclonal antibody Campath 1-H in Chinese hamster ovary cells. In *Animal cell technology: developments, processes and products*, eds. Spier, R.E., Griffiths, J.B. & MacDonald, C. London: Butterworth-Heinemann, pp. 51–3.

Rehemtulla, A., Dorner, A.J. & Kaufman, R.J. (1992), Regulation of PACE propeptide processing activity: requirements for a post-endoplasmic reticulum compartment and autoproteolytic activation. *Proc. Natl. Acad. Sci. USA 89*: 8235–9.

Reiter, M., Hohenwarter, O., Gaida, T., Zach, N., Schmatz, C., Bluml, G., Weigang, F., Nillson, K. & Katinger, H. (1990), The use of macroporous gelatin carriers for the cultivation of mammalian cells in fluidised bed reactors. *Cytotechnology 3*: 271–7.

Reiter, M., Bluml, G., Gaida, T., Zach, N., Unterluggaer, F., Doblhoff-Dier, O., Noe, M., Plail, R., Huss, S. & Katinger, H. (1991), Modular integrated fluidised bed reactor technology. *Bio/Technology 9*: 1100–2.

Reiter, M., Zach, N., Gaida, T., Bluml, G., Doblhoff-Dier, O., Unterluggauer, F. & Katinger, H. (1992a), Oxygenation in fluidised bed bioreactors using the microsparging technique. In *Animal cell technology: developments, processes and products*, eds. Spier, R.E., Griffiths, J.B. & MacDonald, C. London: Butterworth-Heinemann, pp. 386–92.

Reiter, M., Bluml, G., Gaida, T., Zach, N., Schmatz, C., Borth, N., Hohenwarter, O. & Katinger, H. (1992b), High density aggregate cultures of recombinant CHO cells in fluidised bed bioreactors. In *Animal cell technology: developments, processes and products*, eds. Spier, R.E., Griffiths, J.B. & MacDonald, C. London: Butterworth-Heinemann, pp. 421–3.

Reitzer, L.J., Wice, B.M., & Kennell, D. (1979), Evidence that glutamine, not

sugar, is the major energy source for cultured HeLa cells. *J. Biol. Chem. 254*: 2669–76.
Renner, W.A., Jordan, M., & Eppenberger, H.M. (1993), Cell-cell adhesion and aggregates: influence on the growth behaviour of CHO cells. *Biotechnol. Bioeng. 41*: 188–93.
Reuveny, S., Velez, D., MacMillan, J.D., & Miller, L. (1986), Factors affecting growth and monoclonal antibody production in stirred reactors. *J. Immunol. Meth. 86*: 53–9.
Rhodes, M. & Birch, J.R. (1988), Large-scale production of protein from mammalian cells. *Bio/Technology 6*: 518–23.
Rigby, P.W.J. (1982), Expression of cloned genes in eukaryotic cells using vector systems derived from viral replicons. In *Genetic engineering*, Vol. 3., ed. Williamson, R. New York: Academic Press, pp. 84–141.
Rigby P.W.J. (1983), Cloning vectors derived from animal viruses. *J. Gen. Virol. 64*: 255–66.
Robert, J., Côté, J. & Archambault, J. (1992), Surface immobilisation of anchorage dependent mammalian cells. *Biotechnol. Bioeng. 39*: 697–706.
Röder, B. & Kutsch, H. (1992), Collagenated vesicular plant parenchym material as a new type of surface carrier in animal cell culture. In *Animal cell technology: developments, processes and products*, eds. Spier, R.E., Griffiths, J.B. & MacDonald, C. London: Butterworth-Heinemann, pp. 465–8.
Roeder, P.L. & Harkness, J.W. (1986), BVD virus infection: prospects for control. *Vet. Record Feb. 8:* 143–7.
Rosenthal, M.D. (1987), Fatty acid metabolism of isolated mammalian cells. *Prog. Lipid. Res. 26*: 87–93.
Rosevear, A. & Lambe, C (1988), Downstream processing-recent developments. In *Animal cell biotechnology*, eds. Spier, R.E. & Griffiths, J.B., Vol. 3. London: Academic Press, pp. 394–440.
Runstandler, P.W. (1992), Perfusion versus batch culture: which is the preferred process. In *Animal cell technology: developments, processes and products*, eds. Spier, R.E., Griffiths, J.B. & MacDonald, C. London: Butterworth-Heinemann, pp. 347–50.
Runstandler, P.W. & Cernek, S.R. (1987), Large-scale fluidized bed immobilized cultivation of animal cells at high densities. In *Animal cell biotechnology*, eds. Spier, R.E. & Griffiths, J.B. London: Academic Press, pp. 306–320.
Sakai, Y., Furukawa, K. & Suzuki, M. (1992), Immobilisation and long term albumin secretion of hepatocyte spheroids rapidly formed by rotational tissue culture methods. *Biotechnol. Tech. 6*: 517–32.
Sandonini, C.A. & Di Biascio, A. (1992), An investigation of the diffusion limited growth of animal cells around single hollow fibres. *Biotechnol. Bioeng. 40*: 1233–42.
Sarver, N., Gruss, P., Law, M.F., Khoury G. & Hawley, P. (1981), Bovine papilloma virus DNA: a novel eucaryotic cloning vector. *Mol. Cell. Biol. 1*: 486–96.
Scheirer, W., Kanzler, O. & Konopitzky, K.O. (1992), The benefits of using continuous fermentation processes for the production of monoclonal antibodies. *Chimica Oggi:* April, pp. 13–18.
Scheper, T. (1990), Biosensor systems for process control in biotechnology. *Bio. Forum Europe, 7*: 67–70.
Schimke, R. (1978), Gene amplification in animal cells. *Cell 37*: 705–13.

Schlaeger, E.J., Eggimann, B. & Gast, A. (1987), Proteolytic activity in the culture supernatant of mouse hybridoma cells. *Dev. Biol. Stand. 66*: 403–8.

Schleicher, J.B. & Weiss, R.E. (1968), Application of a multiple surface tissue culture propagator for the production of cell monolayers, virus and biochemicals. *Biotech. Bioeng. 10*: 617–24.

Schumpp, B. & Schlaeger, E-J. (1992), Growth study of lactate and ammonia double resistant clones of HL-60 cells. In *Animal cell technology: developments, processes and products*, eds. Spier, R.E., Griffiths, J.B. & MacDonald, C. London: Butterworth-Heinemann, pp. 183–5.

Scholz, M. & Hu, W-S. (1992), A two compartment cell entrapment bioreactor with three different holding times for cells, high and low molecular weight compounds. *Cytotechnology*, in press.

Sekhri, P., DiLeo, A., Alegnezza, A. & Levy, R. (1992), A unique and validatable membrane-based system capable of high resolution removal of virus from protein solutions. In *Animal cell technology: developments, processes and products*, eds. Spier, R.E., Griffiths, J.B. & MacDonald, C. London: Butterworth-Heinemann, pp. 635–8.

Sekiguchi, T., Nishimoto, T., Kai, R. & Sekiguchi, M. (1983), Recovery of a hybrid vector derived from bovine papilloma virus DNA, pBR322 and the HSV *tk* gene by bacterial transformation with extrachromosomal DNA from transfected cells. *Gene 21*: 267–72.

Seo, J.H. & Bailey, J.E. (1985), Effect of recombinant plasmid content on growth properties and cloned gene product formation in *E.coli*. *Biotech. Bioeng. 27*: 1668–74.

Shah, K. & Nathanson, N. (1976), Human response to SV40: review and comment. *Am. J. Epidemiol. 103*: 1–12.

Shuler, M.L., Cho, T., Wickham, T., Ogonah, O., Kool, M., Hammer, D.A., Granados, R.R. & Wood, M.A. (1990), Bioreactor development for production of viral pesticides or heterologous proteins in insect cell cultures. *Ann. N.Y. Acad. Sci. 589*: 399–421.

Siegel, U., Fenge, C. & Fraune, E. (1992), Spin filter for continuous perfusion of suspension cells. In *Animal cell technology: developments, processes and products*, eds. Spier, R.E., Griffiths, J.B. & MacDonald, C. London: Butterworth-Heinemann, pp. 434–6.

Sisson, J. & Ellis, L. (1989), Secretion of the extracellular domain of the human insulin receptor from insect cells by use of a baculovirus vector. *Biochem J. 261*: 119–26.

Smeekens, S.P. (1993), Processing of protein precursors by a novel family of subtilisin-related mammalian endoproteases. *Bio/Technology 11*: 182–6.

Smith, G.L. (1991), Vaccinia virus vectors for gene expression. *Current Opinion Biotech. 2*: 713–17.

Smith, G.E., Ju, G., Ericson, B.L., Moshera, J., Latim, H-W., Chizzonite, R. & Summers, M.D. (1985), Modification and secretion of human interleukin 2 produced in insect cells by a baculovirus expression vector. *Proc. Natl. Acad. Sci. USA 82*: 8404–8.

Smith, G.L., Mackett, M. & Moss, B. (1983), Infectious vaccinia virus recombinants that express hepatitis B virus surface antigen. *Nature 302*: 490–5.

Smith, G.L, Murphy, B.R. & Moss, B. (1983), Construction and characterisation of an infectious vaccinia virus recombinant that expresses the influenza haemagglutinin gene and induces resistance to influenza virus in hamsters. *Proc. Natl. Acad. Sci. USA 80*: 7155–9.

Smith, G.E., Summers, M.D. & Fraser, M.J. (1983), Production of human beta-interferons in insect cells infected with a baculovirus expression vector. *Mol. Cell. Biol. 3*: 2156–63.

Smith, K.T., Doherty, I., Thomas, J.A., Per, S.R. & Sito, A.F. (1992), Quantitation of residual DNA in biological products: new regulatory concerns and new methodologies. In *Animal cell technology: developments, processes and products*, eds. Spier, R.E., Griffiths, J.B. & Meignier, B. London: Butterworth-Heinemann, pp. 696–8.

Spier, R.E. (1992), Animal cell biotechnology in the 1990's: from models to morals. In *Animal cell biotechnology,* Vol. 5, eds. Spier, R.E. & Griffiths, J.B. London: Academic Press, pp. 1–46.

Spier, R.E. & Whiteside, J.P. (1976), The production of foot and mouth disease virus from BHK21 cells grown on the surface of glass spheres. *Biotechnol. Bioeng. 18*: 649–57.

Stacey, G.N., Bolton, B.J., Morgan, D. & Doyle, P.G. (1992), Validation of DNA fingerprinting in animal cell technology: the differentiation and identification of murine hybridoma clones from a single fusion and the stability of the transformed cell lines. In *Animal cell technology: developments, processes and products*, eds. Spier, R.E., Griffiths, J.B. & MacDonald, C. London: Butterworth-Heinemann, pp. 681–7.

Stark, G. & Wahl, G. (1984), Gene amplification. *Ann. Rev. Biochem. 53*: 447–91.

Stark, N.J. & Heath, E.C. (1979), Glucose-dependent glycosylation of secretory glycoprotein in mouse myeloma cells. *Arch. Biochem. Biophys. 192*: 599–609.

Stenlund, A., Lamy, D., Moreno-Lopez, J., Ahola, H., Pettersson, V. & Tiollais, P. (1983), Secretion of hepatitis B virus surface antigen from mouse cells using an extra-chromosomal eucaryotic vector. *Embo J. 2*: 669–73.

Stephens, P.E. & Hentshel, C.C.G. (1987), The bovine papilloma virus genome and its use as a eukaryotic vector. *Biochem J. 248*: 1–11.

Strasser, A., Whittingham, S., Vaux, D.L., Bath, M.L., Adams, J.M., Cory, S. & Harris, A.W. (1991), Enforced BCL2 expression in B-lymphoid cells prolongs antibody responses and elicits auto-immune disease. *Proc. Nat. Acad. Sci. USA 88*: 8661–5.

Sugden, B., Marsh, K., & Yates, J. (1985), A vector which replicates as a plasmid and can be efficiently selected in B-lymphocytes transformed by Epstein-Barr virus. *Mol. Cell. Biol. 5*: 410–13.

Sugimoto, S., Lind, W. & Wagner, R. (1992), Activation of a specific proteolytic activity in suspension cultures of recombinant adherent cells. In *Animal cell technology: developments, processes and products*, eds. Spier, R.E., Griffiths, J.B. & MacDonald, C. London: Butterworth-Heinemann, pp. 547–51.

Sugiura, T. (1992), Effects of glucose on the production of recombinant protein C in mammalian cell culture. *Biotechnol. Bioeng. 39*: 953–9.

Sumeghy, Z. (1992), Improved cell retention system based on the rotating sieve technique. In *Animal cell technology: developments, processes and products*, eds. Spier, R.E., Griffiths, J.B. & Meignier, B. London: Butterworth-Heinemann, pp. 437–46.

Suput, D. (1984), The effect of external ammonium on the kinetics of the sodium current in frog muscle. *Biochem. Biophys. Acta. 771*: 1–8.

Taggart, R.T. & Samloff, I.M. (1983), Stable antibody-producing hybridomas. *Science 219*: 1228–30.

Tagler, J.M., Ackerboom, T.P.M., Hoek, J.B., Meijer, A.J., Vaartjes, W., Ernster, L. & Williamson, J.R. (1975), Ammonia and energy metabolism in isolated mitochondria and intact liver cells. In *Normal and pathological development of energy metabolism*, ed. Hommes, F.A. & van der Berg, C.J. London: Academic Press, pp. 63–75.

Takehara, K., Ireland, D. & Bishop, D.H.L. (1988), Co-expression of the hepatitis B surface and core antigens using baculovirus multiple expression vectors. *J. Gen. Virol. 69*: 2763–77.

Talbot, D., Collis, P., Antoniou, M., Vidal, M., Grosveld, F. & Greaves, D.R. (1989), A dominant control region from the human beta-globin locus conferring integration site independent gene expression. *Nature 338*: 352–5.

Telling, R.C. & Radlett, P.J. (1970), Large scale cultivation of mammalian cells. *Adv. Appl. Microbiol. 13*: 91–119.

Thacker, J., Webb, M.B.T. & Debenham, P.G. (1988), Fingerprinting cell lines: use of human hypervariable DNA probes to characterise mammalian cell cultures. *Som. Cell. Mol. Gen. 14*: 519–25.

Thaler, T. & Varnak, J. (1991), Oxygen transfer characteristics in cell culture fermentors: direct sparging, membrane oxygenation, bubble free aeration through a rotating sieve. In *Production of biologicals from animal cells in culture*, eds. Spier, R.E., Griffiths, J.B. & Meignier, B. London: Butterworth-Heinemann, pp. 451–3.

Tharakan, J.P. & Chau, P.C. (1986), Serum free fed batch production of IgM. *Biotech. Lett. 8*: 457–62.

Thomas, R.H., Jenkins, H.A. & Butler, M. (1991), Adaptation of anchorage dependent cells to glutamine-free medium. In *Production of biologicals from animal cells in culture*, eds. Spier, R.E., Griffiths, J.B. & Meignier, B. London: Butterworth-Heinemann, pp. 276–8.

Thommes, J., Garske, U., Biselli, M. & Wandrey, C. (1992), Integrated detoxification: reduction of ammonium concentration by dialysis with cation exchange membranes. In *Animal cell technology: developments, processes and products*, eds. Spier, R.E., Griffiths, J.B. & MacDonald, C. London: Butterworth-Heinemann, pp. 171–5.

Thorens, B. & Vassali, P. (1986), Chloroquine and ammonium chloride prevent terminal glycoylation of immunoglobulins in plasma cells without affecting secretion. *Nature 321*: 53–62.

Tokashiki, M., Hamemoto, K., Takazawa, Y. & Ichikawa, Y. (1988), High density culture of mouse-human hybridoma cells using a new perfusion culture vessel. *Kagakukogaku Ronbunshu 14*: 337–41.

Tokashiki, M., Arai, T., Hamamoto, K. & Ishimara, K. (1990), High density culture of hybridoma cells using a perfusion culture vessel with an external centrifuge. *Cytotechnology 3*:239–44.

Tokashiki, M.& Arai, T. (1991), High density culture of hybridoma cells using perfusion culture apparatus with multi-settling zones. In *Production of biologicals from animal cells in culture*, eds. Spier, R.E., Griffiths, J.B. & Meignier, B. London: Butterworth-Heinemann, pp. 467–9.

Traunecker, A., Oliveri, F. & Karjalainen, K. (1991), Myeloma based expression system for production of large mammalian proteins. *Tibtech 9*: 109–13.

Tso, E.I., Bohn, M.A., Omstead, D.R. & Munster, M.J. (1991), Optimisation of a roller bottle process for the production of recombinant erythropoietin. *Ann. N. Y. Acad. Sci. 665*: 127–36.

Urlaub, G. & Chasin, L. (1980), Isolation of Chinese hamster cell mutants deficient in dihydrofolate reductase activity. *Proc. Natl. Acad. Sci. USA 77:* 1271–4.

Van der Marel, P. (1985), Concentration. In *Animal cell biotechnology,* Vol. 2, eds. Spier, R.E. & Griffiths, J.B. London: Academic Press, pp. 185–215.

Van der Pol, L. & Tramper, J. (1992), Effect of reducing the serum or albumin concentrations on the shear sensitivity of two hybridoma cell lines in sparged cultures. In *Animal cell technology: developments, processes and products*, eds. Spier, R.E., Griffiths, J.B. & MacDonald, C. London: Butterworth-Heinemann, pp. 192–4.

Van der Straten, A., Johansen, H., Sweet, R. & Rosenberg, M. (1989), Efficient expression of foreign genes in cultured *Drosophila melanogoster* cells using hygromycin B selection. In *Invertebrate cell system applications*, Vol. 1, ed. Mitsuhashi, J. Boca Raton, Florida: CRC Press Inc., pp. 183–95.

Van Wezel, A.L. (1967), Growth of cell strains and primary cells on microcarriers in homogeneous culture. *Nature 216*: 64–5.

Van Zjil, P.C.M., Moonen, C.T., Faustino, P., Pekar, J., Kaplan, O. & Cohen, J.S. (1991), Complete separation of intracellular and extracellular information in NMR spectra of perfused cells by diffusion-weighted spectroscopy. *Proc. Natl. Acad. Sci. USA 88*: 3228–32.

Varecka, R. & Schierer, W. (1987), Use of a rotating wire cage for the retention of animal cells in a perfusion fermenter. *Develop. Biol. Standard. 66*: 269–72.

Vialard, J.E., Lalumiere, M., Vernet, T. Briedis, D., Alkhatib, G., Henning, D., Levin, D. & Richardson, C. (1990), Synthesis of the membrane fusion and haemagglutination proteins of measles virus using a novel baculovirus vector containing the B-galactosidase gene. *J. Virol. 64*: 5804–11.

Vidal, M., Wrighton, C., Eccles, S., Burke, J. & Grosveld, F. (1990), Differences in human cells to support stable replication of Epstein-Barr virus-based shuttle vectors. *Biochim. Biophys. Acta 1048*: 171–7.

Walter, J.K., Werz, W. & Berthold, W. (1992), Virus removal and inactivation: concept and data for process validation of downstream processing. *Biotech. For. Europe 9*: 560–4.

Walter, J., Werz, W., McGoff, P., Werner, R-G. & Berthold, W. (1992), Virus removal/inactivation in down stream processing. In *Animal cell technology: developments, processes and products*, eds Spier, R.E., Griffiths, J.B. & MacDonald, C. London: Butterworth-Heinemann, pp. 624–34.

Wang, X.C., O'Hanlon, T.P. & Lau, J.T.Y. (1989), Regulation of a-galactoside ' 2.6 sialyltransferase gene expression by dexamethasone. *J. Biol. Chem. 264*: 1854–9.

Wang, G., Zhang, W., Freedman, D., Eppstein, L. & Kadouri, A. (1992), Continous production of monoclonal antibodies in Celligen packed bed reactor using Fibra-cell carrier. In *Animal cell technology: developments, processes and products*, eds. Spier, R.E., Griffiths, J.B. & MacDonald, C. London: Butterworth-Heinemann, pp. 460–3.

Wegner, M., Zastrow, G., Klavinius, A., Schwender, S., Muller, F., Luksza, H., Hoppe, J., Wrenburg, J. & Grummt, F. (1989), Cis-acting sequences from mouse rDNA promote plasmid DNA amplification and persistence in mouse cells: implications of HMG-7 in their function. *Nuc. Acid. Res 17*: 9909–32.

Weidle, U.H., Buckel, P. & Weinburg, J. (1988), *Gene 66*: 193–203.

Weidle, U.H. & Buckel, P. (1987), Establishment of stable mouse myeloma cells constitutively secreting human tPA. *Gene 57*: 131–41.

Weiss, R.E. & Schleicher, J.B. (1968), A multi-surface tissue propagator for the mass-scale growth of cell monolayers. *Biotech. Bioeng. 10*: 601–15.

Welply, J.K. (1989), Sequencing methods for carbohydrates and their biological applications. *Tibtech 7*: 5–10.

White, A.G., Raju, K., Keddie, S. & Abouna, G.M. (1989), Lymphocyte activation: changes in intracellular adenosine triphosphate and DNA synthesis. *Immunol. Lett. 22*: 47–50.

Whittle, N., Adair, J. & Lloyd, C. (1987), Expression in COS cells of a mouse-human chimaeric B72.3 antibody. *Protein Engineering 81*: 3806–10.

Williams, G.T., Smith, C.A., Spooncer, E., Dexter, T.M. & Taylor, D.R. (1990), Haematopoietic colony stimulating factors promote cell survival by suppressing apoptosis. *Nature 343*: 76–9.

Wood, C.R., Doner, A.J., Morris, G.E., Alderman, E.M., Wilson, D., O'Hara, R.M. & Kaufman, R.J. (1990), High level synthesis of immunoglobulins in Chinese hamster ovary cells. *J. Immunol. 145*: 3011–16.

Wu, P., Ray, G. & Shuler, M.L. (1991), A single cell model for CHO cells. *Ann. N.Y. Acad. Sci. 665*: 152–87.

Wurm, F.M., Johnson, A., Lie, Y., Etcheverry, T. & Anderson, K.P. (1992), Host cell derived retroviral sequences enhance transfection and expression facility in CHO cells. In *Animal cell technology: developments, processes and products*, eds. Spier, R.E., Griffiths, J.B. & MacDonald, C. London: Butterworth-Heinemann, pp. 35–41.

Yamane, I., Kan, M., Minamoto, Y. & Amatsuji, Y. (1982), Alpha-cyclodextrin: a partial substitute for bovine serum albumin in serum free culture of mammalian cells. In *Growth of cells in hormonally defined media*, eds. Sato, G.H., Pardee, A.B. & Sirbaska, D.A. Cold Spring Harbour NY: Cold Spring Harbour Conferences on Cell Proliferation, pp. 87–92.

Yanagi, H., Ogawa, I., Hozumi, T., Okamoto, M. & Yoshima, T. (1989), High level expression of human erythropoietin cDNA in stably transfected Namalwa cells. *J. Ferment. Bioeng. 68*: 257–63.

Yates, J., Warren, N. & Sugden, B. (1985), Stable replication of plasmids derived from Epstein-Barr virus in mammalian cells. *Nature 313*: 812–15.

Yates, J., Warren, N., Reisman, D. & Sugden, B. (1984), A cis-acting element from the Epstein-Barr viral genome that permits stable replication of recombinant plasmids in latently infected cells. *Proc. Natl. Acad. Sci. USA 81*: 3806–10.

Zastrow, G., Koehler, U., Muller, F., Klavinius, A., Wegner, M., Wienbury, J., Weidle, U.H. & Grummt, F. (1989), Distinct mouse DNA sequences enable establishment and persistence of plasmid DNA polymers in mouse cells. *Nuc. Acid. Res. 17*: 1867–79.

Zheng, H. & Wilson, J.H. (1990), Gene targeting in normal and amplified cell lines. *Nature 344*: 170–3.

Zielke, H.R., Ozand, P.T., Tildon, J.T., Sevdahan, D.A. & Cornblath, M. (1978), Reciprocal regulation of glucose and glutamine utilisation by cultured human diploid fibroblasts. *J. Cell. Physiol. 95*: 41–8.

Zinn, K., Mellon, P., Ptashne, M. & Maniates, T. (1982), Regulated expression of an extrachromosomal beta-interferon gene in mouse cells. *Proc. Natl. Acad. Sci. USA 79:* 4897–4901.

Zu Pulitz, J., Kubasek, W.K., Duchene, M., Marget, M., von Specht, B-U. & Domdey, H. (1990), Antibody production in baculovirus-infected insect cells. *Bio/Technology 8:* 651–4.

Index